LOW Pressure BOILERS

second edition

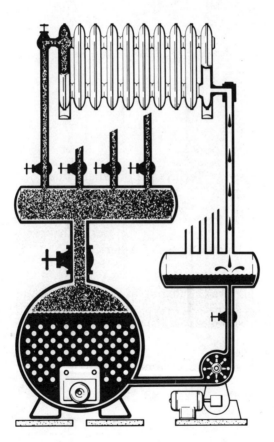

Frederick M. Steingress

AMERICAN TECHNICAL PUBLISHERS, INC.
HOMEWOOD, ILLINOIS 60430

The author and publisher are grateful for the technical information and assistance provided by the following companies.

Bell & Gossett Co., ITT Corp.
Big Three Industries,
  Tempil Division
Cleaver-Brooks
Combustion Engineering Co.
Electronic Control
Honeywell, Inc.
Jenkins Bros.
Kennedy Valve Co.

The Kraissl Co. Inc.
Lunkenheimer Co.
McDonnell & Miller, Inc.
Manning, Maxwell,
  and Moore, Inc.
The Permutit Co. Inc.
Sarco
Walter Kidde

2 3 4 5 6 7 8 9 -86- 9 8

Printed in the United States of America

Library of Congress Cataloging-in-Publication Data
Steingress, Frederick M.
  Low pressure boilers.

  Includes index.
  1. Steam-boilers.   2. Steam-heating, Low pressure.
I. Title.
TJ286.S73 1986   621.1′83   86–10865
ISBN 0–8269–4407–8 (pbk.)

CONTENTS

INTRODUCTION

LOW PRESSURE BOILERS, 2nd Edition provides information on the safe and efficient operation of low pressure boilers and related equipment. The book can be used as an introduction to stationary engineering or as a reference book for upgrading skills. Boiler operation and boiler components are broken down into separate topics in the book. In the 2nd edition, a new chapter entitled Hot Water Heating Systems has been added.

At the beginning of each chapter, arrowhead symbols are shown that are used in the chapter to identify the flow of material in boiler equipment or a boiler system. As a material changes, the arrowhead symbol also changes. For example, water when heated turns to steam and steam when cooled turns to condensate.

KEY TO ARROWHEAD SYMBOLS			
⟁ - Air	▲ - Water	▲ - Fuel Oil	⌂ - Air to Atmosphere
⟁ - Gas	△ - Steam	△ - Condensate	⌂ - Gases of Combustion

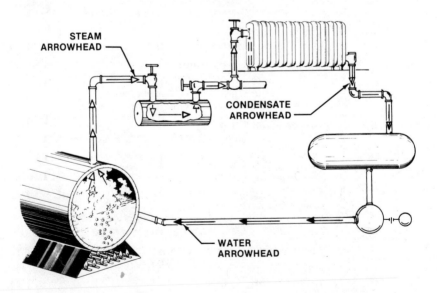

Points to remember in the chapters identify pertinent information previously covered. Key words that are defined in the chapter are listed at the end of each chapter. The illustrated glossary and the index are provided for easy reference.

THE BOILER 1

Boilers are used for heating buildings and for various industrial processes. A boiler is a closed vessel containing water. The water in the boiler is heated and pressurized when heat is added. The steam is

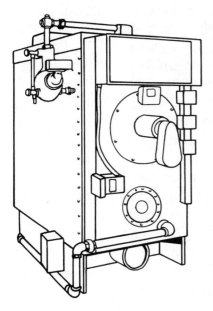

then directed to different locations for use. The water in a hot water boiler is heated and directed to heating units as required. Different accessories are required to supply water, air, and fuel for boiler operation.

Water lost in the steam cycle must be replaced. Air supplied for the burning of fuel in the furnace must be regulated. The correct amount of fuel supplied to the burner is necessary to maintain heat in the boiler.

KEY TO ARROWHEAD SYMBOLS			
△ - Air	▲ - Water	▲ - Fuel Oil	⌂ - Air to Atmosphere
△ - Gas	△ - Steam	△ - Condensate	⌂ - Gases of Combustion

PRINCIPLES OF BOILER OPERATION

A boiler is used to generate steam for heat or for power. To generate steam a container, water, and heat are required. The boiler container serves three purposes. It holds the water, transfers heat to the water to make steam, and collects the steam that is produced.

Water turns to steam when it is heated. A sufficient supply of water is necessary to operate the boiler. *NOTE:* It takes one pound of water to make one pound of steam. Water is used for steam because it is plentiful and inexpensive.

Heat is required to change water to steam. Heat can be supplied by the sun, electricity, gas, fuel oil, coal, and wood. Fuel oil, coal, or gas are most commonly used when a large volume of steam is needed for power and heating.

The modern boiler is designed for efficiency of operation. Over the years the boiler design was changed to increase efficiency. See Figure 1-1. A container is filled half full of water, and a fire is built under it. When a lid is added to the container, a simple boiler is formed.

The fire heats the water to about 212 °F and the water begins to boil and turns to steam. To use this steam, it is collected and lead to its working station. A pipe is put on top of the lid of the container so the steam flows up through the pipe and leads the steam to where it is needed.

This arrangement is acceptable but not very efficient. The boiler must be efficient to minimize the cost of producing steam. Boiler efficiency is increased by producing more steam using the same amount of fuel. One way a boiler can be made more efficient is by increasing the *heating surface* (that part of the boiler with water on one side and heat on the other) so that more heat is transferred to the water to produce steam.

The heating surface can be increased and more water put closer to the heat by laying the boiler on its side. The scotch marine boiler uses a large heating surface to transfer heat from the burning fuel. See Figure 1-2.

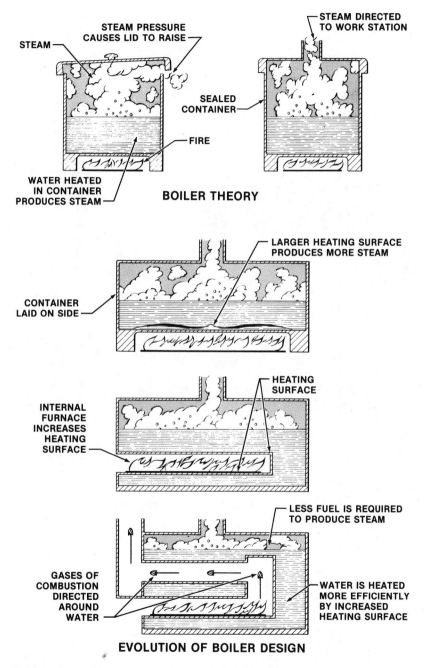

Figure 1-1. More steam is produced by increasing the heating surface in a boiler.

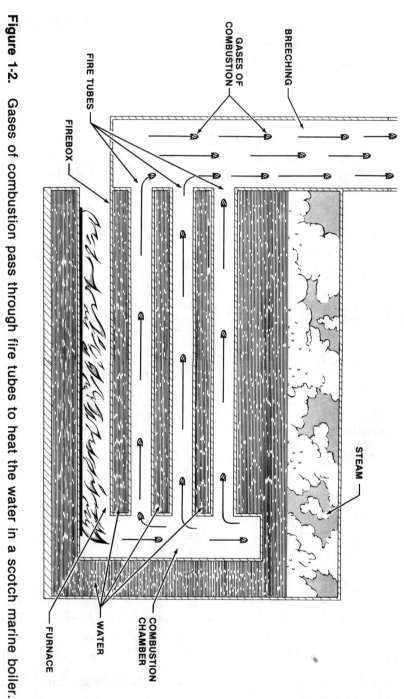

Figure 1-2. Gases of combustion pass through fire tubes to heat the water in a scotch marine boiler.

Heat produced in the furnace is routed through fire tubes where heat is tranferred to the surrounding water. The heated water produces steam. The gases of combustion continue to flow through the fire tubes up into the breeching.

The fire inside the combustion chamber in the boiler allows fuel to burn more efficiently because fuel mixes with a greater quantity of air. Enlarging the combustion chamber increases the quantity of air that can mix with the fuel, allowing fuel to burn at a faster rate. It is still possible to make the boiler work even more efficiently by increasing the size of the heating surface.

The container, water, and heat are still present. The only difference is that more water has been put next to the heated metal, resulting in increased heat transfer. This produces more steam at a lower cost.

Although the boiler is a single unit, four separate but interrelated systems are necessary for operation. These basic systems apply to all boilers regardless of their size or use. The systems are:

1. *feedwater system*—supplies water to the boiler
2. *fuel system*—supplies fuel for making heat
3. *draft system*—provides air for combustion
4. *steam system*—collects and controls the steam that is made

Each of the four systems is necessary to operate a boiler, yet each is an independent system.

Feedwater System

The function of the feedwater system is to feed water to the boiler. See Figure 1-3. Water in the boiler is heated and turns to steam. The steam leaves the boiler by a pipe called the main steam line (boiler outlet) (1) where it enters the main header (2). From the main header, main branch lines (3) carry the steam to a riser (4) and then to the steam heating equipment (5).

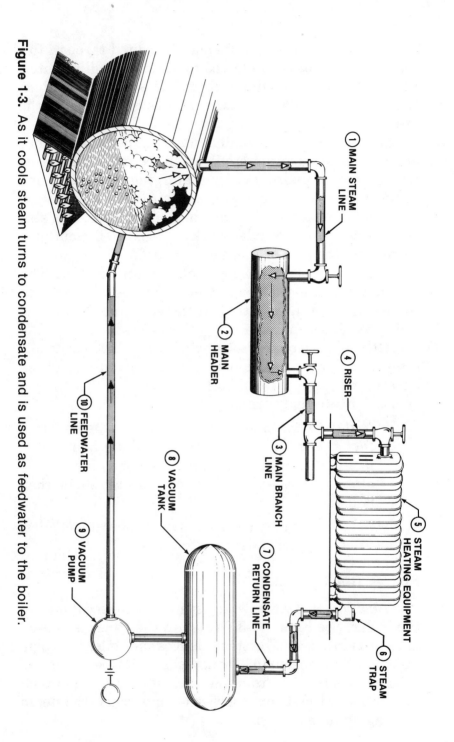

Figure 1-3. As it cools steam turns to condensate and is used as feedwater to the boiler.

At this point the steam in the heating unit cools and turns to water called condensate. The condensate is separated from the heating equipment by a steam trap (6) that allows condensate, but not steam, to pass through. The condensate passes along a pipe called a condensate return line (7) to a vacuum tank (8).

A vacuum pump (9) creates a vacuum that helps to draw the water out of the condensate return line and into the vacuum tank. The vacuum pump also delivers the water or condensate back to the boiler through a pipe called a feedwater line (10). Once it has returned to the boiler, the water again is turned into steam and the process repeats itself.

Fuel System

A fuel system provides fuel to maintain combustion, which generates heat. The three fuels commonly used to heat the water in a low pressure boiler are fuel oil, gas, and coal. The type of fuel used determines the equipment required to store, transport, and burn the fuel. All fuels are combustible and can be dangerous if necessary safety precautions are not followed. The fuel used in a low pressure boiler depends on price and availability.

Fuel Oil System.   In a low pressure boiler fuel oil system, fuel oil is stored in the fuel oil tank. See Figure 1-4. The fuel oil tank is usually buried in the ground. The fuel oil leaves the tank through the suction line and duplex strainer and goes to the fuel oil pump. The fuel oil is forced from the pump under pressure through the discharge line.

From the discharge line, some of the fuel oil goes to the fuel oil burner where it is burned. The rest of the fuel oil goes back to the fuel oil tank through the fuel oil regulating valve to the fuel oil return line.

Gas System.   In the low pressure boiler gas system the same gas burned at home for cooking and heating is used.

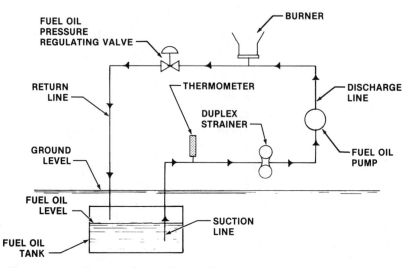

Figure 1-4. Excess fuel oil supplied to the burner is returned to the fuel oil tank.

NOTE: Gas is very toxic and explosive and must be handled with caution. The difference lies in controlling how the gas is burned. In a basic gas system, city gas lines supply gas to the boiler at a set pressure. This pressure varies depending on the city and the particular gas plant.

The two types of gas systems are the *low pressure gas system* and the *high pressure gas system.* In the low pressure gas system, gas is reduced from city gas pressure to 0 pressure. See Figure 1-5. The regulator (1) reduces the city gas pressure to 0 pressure. Leaving the regulator, the gas is drawn into the burner and mixed with air supplied by a blower (2).

This mixture then passes through the burner (3) into the furnace where it is ignited by a continuously burning pilot (4). The gases of combustion leave the boiler through the breeching to the chimney.

In a high pressure gas system, gas is supplied to the boiler room from a high pressure gas main and is at a

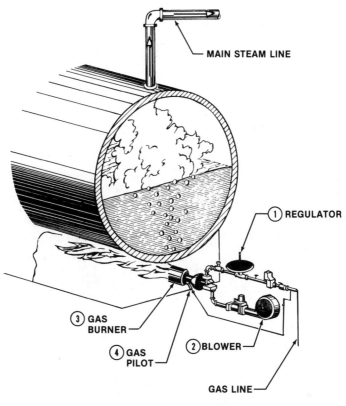

MAIN STEAM LINE

① REGULATOR

③ GAS BURNER

④ GAS PILOT

② BLOWER

GAS LINE

LOW PRESSURE GAS SYSTEM

Figure 1-5. Gas pressure is reduced to 0 psi in the low pressure gas system.

higher pressure than the gas used in a home. See Figure 1-6. In a high pressure gas system, the gas line is fitted with a manual shutoff gas cock (1) that allows the gas to be secured when working on the gas lines or in the boiler furnace.

The gas then passes through the gas pressure regulator (2), which reduces the incoming gas to the required pressure at the burner. The gas pressure gauge (3) located after the gas pressure regulator indicates the pressure of

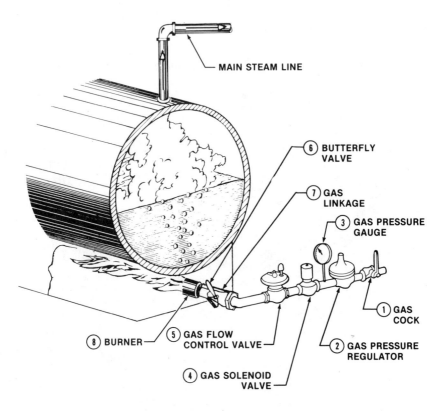

MAIN STEAM LINE

⑥ BUTTERFLY VALVE

⑦ GAS LINKAGE

③ GAS PRESSURE GAUGE

① GAS COCK

⑧ BURNER

⑤ GAS FLOW CONTROL VALVE

② GAS PRESSURE REGULATOR

④ GAS SOLENOID VALVE

HIGH PRESSURE GAS SYSTEM

Figure 1-6. The gas pressure regulator reduces gas pressure to the required pressure at the burner in a high pressure gas system.

gas being supplied to the burner. The gas solenoid valve (electric valve) (4) opens at the proper time in the cycle to admit gas to the gas flow control valve (5).

The gas flow control valve only allows a very small amount of gas to pass through until ignition is established and then slowly opens to allow a maximum flow of gas to the butterfly valve (6). The butterfly valve controls the firing rate through the gas linkage (7).

Boilers may burn gas only or have a combination

gas/fuel oil burner. The gas lines and equipment on a combination gas/fuel oil burner are the same up to the burner in both boilers.

Coal System. In a coal system, coal is fed to the boiler by hand firing or by a coal stoker. Hand firing is accomplished by manually shoveling coal into the furnace. Hand firing is inefficient because the fire doors must be left open while the coal is being shoveled. This results in a loss of heat and cooling of the furnace and brickwork, reducing efficiency. Hand firing is rarely used in modern low pressure plants.

Stoker firing is a mechanical method of feeding coal to the furnace. See Figure 1-7. Coal stokers are more effi-

Figure 1-7. Coal stokers feed coal mechanically to the furnace at a consistent rate. (*Combustion Engineering Co.*)

cient than hand firing. At one time, coal was the most popular fuel used to fire steam boilers. However, fuel oil and gas have become more popular due to price, availability, and pollution standards.

Draft System

The draft system provides and regulates air to the burner in a low pressure boiler. Fuels require oxygen in air to burn. To burn fuel the correct amount of air must be provided to the boiler furnace. In addition, air must pass into the furnace and be allowed to leave the furnace in the form of gases of combustion after it has mixed with the fuel and burned. See Figure 1-8.

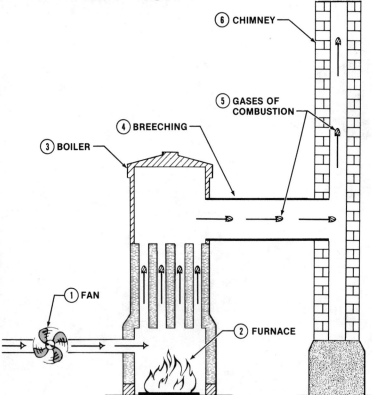

Figure 1-8. The draft system provides air for combustion to the burner and removes gases of combustion from the burner.

A fan (1) draws air in and supplies it to the boiler furnace (2) at slight pressure. In the boiler furnace the air mixes with the fuel and burns. The gases of combustion pass through the boiler furnace around the boiler (3) and enter the breeching (smoke pipe) (4). From the breeching the gases of combustion (5) enter a chimney (stack) (6) and are released into the atmosphere.

Steam System

The function of a boiler is to produce steam. Steam produced by heating water must be controlled and transported to the locations where it is to be used. The steam system shares some of the same components with the water system.

The function of the steam system is to remove air, control steam flow, and maintain the required steam pressure. The steam system includes vents, valves, headers, piping, safety valves, and radiators or heating units.

Points to Remember

1. The three factors necessary to produce steam are a container, water, and heat.
2. The container holds water, transfers heat to the water, and collects the steam produced.
3. Water is supplied to the boiler by a feedwater system that usually collects condensed steam and returns it.
4. Heat is supplied by burning some type of fuel in a combustion chamber. Fuel oil, gas, and coal are the fuels used in low pressure boilers.
5. A draft system is needed to supply the necessary air for combustion to the furnace.

BOILER TYPES

Low pressure boiler types vary to suit different load requirements. The basic principles on which boilers operate are the same regardless of type. Low pressure steam boilers are used primarily for heating buildings such as schools, apartments, warehouses, and factories, and for heating domestic water. Boiler size will vary based on the quantity of steam required.

A low pressure steam boiler has a maximum allowable working pressure (MAWP) of 15 pounds per square inch (psi). This may vary in some states. Check your local pressure vessel code to determine the maximum psi allowed for low pressure steam boilers. The three basic types of low pressure steam boilers are the *fire tube boiler*, the *water tube boiler*, and the *cast iron sectional boiler*. See Figure 1-9.

Fire Tube Boiler

In a fire tube boiler, heat and gases of combustion pass through fire tubes that are surrounded by water. Fire tube boilers include the *scotch marine boiler*, the *firebox boiler*, and the *vertical fire tube boiler*. See Figure 1-10. The scotch marine boiler is long, low, and round. The firebox boiler has two different lengths of tubes and requires more brick-work than the scotch marine boiler.

The vertical fire tube boiler is no longer used in large installations but is still popular for residential use. The two types of vertical fire tube boilers are the *dry-top* and the *wet-top*. In the dry-top vertical fire tube boiler the upper portion of the fire tubes is exposed to the gases of combustion. In the wet-top vertical fire tube boiler the upper portion of the fire tubes is submerged in water.

Water Tube Boiler

In a water tube boiler, the water is inside the tubes and gases of combustion flow around the tubes. The most common type of water tube boiler is the *straight tube multiple-pass boiler*. See Figure 1-11. Gases of combustion pass around the tubes three times before discharging to the

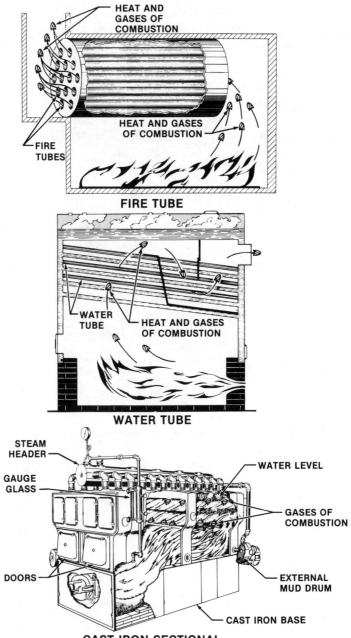

FIRE TUBE

WATER TUBE

CAST IRON SECTIONAL

Figure 1-9. Low pressure boilers have a maximum allowable working pressure (MAWP) of 15 psi and can be fire tube, water tube, or cast iron sectional.

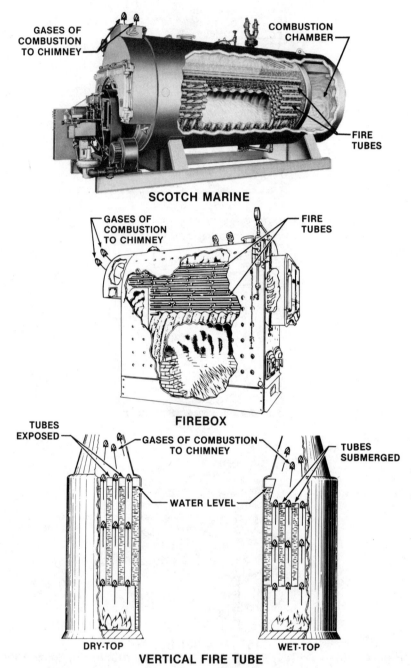

GASES OF
COMBUSTION
TO CHIMNEY

COMBUSTION
CHAMBER

FIRE
TUBES

SCOTCH MARINE

GASES OF
COMBUSTION
TO CHIMNEY

FIRE
TUBES

FIREBOX

TUBES
EXPOSED

GASES OF COMBUSTION
TO CHIMNEY

TUBES
SUBMERGED

WATER LEVEL

DRY-TOP

WET-TOP

VERTICAL FIRE TUBE

Figure 1-10. Heat and gases of combustion pass through fire tubes surrounded by water in a fire tube boiler.

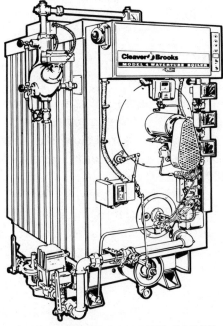

WATER TUBE HOT WATER BOILER

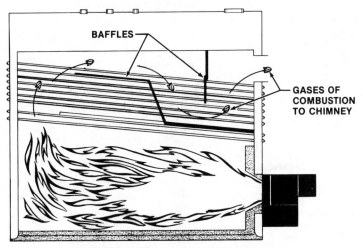

WATER TUBE STEAM BOILER

Figure 1-11. Water passes through water tubes surrounded by heat and gases of combustion in a water tube boiler. (*Cleaver-Brooks*)

chimney. This provides maximum heat transfer from the gases of combustion to the water in the tubes.

Cast Iron Sectional Boiler

In a cast iron sectional boiler, gases of combustion flow around sections that contain water. These boilers are sometimes referred to as water tube cast iron boilers. However, there are no tubes in a cast iron sectional boiler. Cast iron sections containing water are joined together to form a boiler. See Figure 1-12.

The cast iron sectional boiler may have 5 sections for a small building and 12 sections for a large building. The sections are fastened together to provide the size needed. Cast iron sectional boilers are sometimes called pork chop boilers because the individual sections resemble a pork chop in shape.

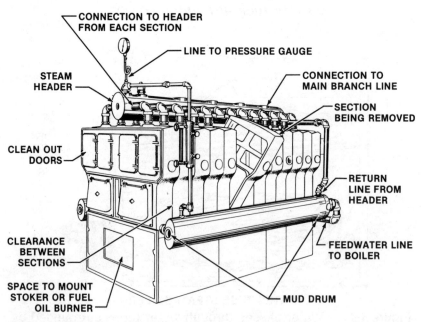

Figure 1-12. The capacity of a cast iron sectional boiler can be increased by adding sections to the boiler.

Points to Remember ————————————

1. Boiler size and type vary depending upon the load requirements.
2. The three basic types of boilers are fire tube, water tube, and cast iron sectional.
3. Fire tube boilers have the heat from the fire and the gases of combustion inside the tubes.
4. The three types of fire tube boilers are the scotch marine, firebox, and vertical fire tube boiler.
5. The firebox boiler requires more brickwork than the scotch marine boiler.
6. The vertical fire tube boiler is used primarily in residential applications.
7. Water tube boilers have the water inside the tubes and the heat on the outside.
8. The most common type of water tube boiler is the straight tube multiple-pass boiler.
9. Straight tube multiple-pass boilers use baffles to direct gases of combustion around the water tubes for maximum heat transfer.
10. Cast iron sectional boilers have large sections with water inside and heat flowing around the sections.
11. Cast iron sectional boilers can be expanded by adding sections to the boiler.
12. Low pressure steam boilers have a maximum allowable working pressure (MAWP) of 15 pounds per square inch (psi).

Key Words _____

boiler
breeching
cast iron sectional boiler
coal system
combustion chamber
condensate
draft system
feedwater system
firebox boiler
fire tube boiler
fuel oil system

fuel system
gas system
hand firing
heating surface
scotch marine boiler
steam
steam system
stoker firing
straight-tube multiple
  pass boiler
water tube boiler

BOILER
FITTINGS

2

All boilers, regardless of type, require boiler fittings to operate safely and efficiently. Boiler fittings are located on the boiler for function and accessibility. The ASME (American Society of Mechanical Engineers) code requires that all boiler fittings be constructed of materials that will withstand the temperatures and pressures the boiler will be subjected to. In addition, the operating and testing procedures of boiler fittings are specified by the ASME code based on operating temperatures and pressures of the boiler. The most important boiler fitting is the safety valve. All required boiler fittings have a purpose and must be properly maintained to function properly.

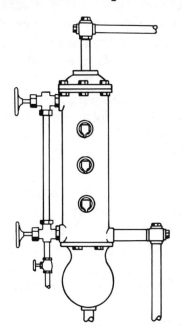

KEY TO ARROWHEAD SYMBOLS			
△ - Air	▲ - Water	▲ - Fuel Oil	⌂ - Air to Atmosphere
⚠ - Gas	△ - Steam	⚠ - Condensate	⚠ - Gases of Combustion

SAFETY VALVE

The purpose of the safety valve is to prevent the pressure in the boiler from exceeding its maximum allowable working pressure (MAWP). Most boiler inspectors feel that the safety valve is the most important valve on a boiler. The MAWP for low pressure boilers is 15 pounds per square inch (psi). The pressure of the steam in the boiler must not exceed 15 psi.

If the pressure were to rise above 15 psi, the safety valve would pop open, preventing any buildup of pressure that might lead to a boiler explosion. The safety valve is located at the highest part of the steam side of the boiler and connected directly to the shell. There must not be any valves located between the safety valve and the boiler.

The capacity of a safety valve is measured in the amount of steam that can be discharged per hour. The safety valve capacity is identified on the data plate attached to the safety valve. For example, if the capacity is listed at 6,900 pounds, the safety valve can discharge 6,900 pounds of steam per hour.

Safety valves are designed to pop open when pressure exceeds the MAWP in the boiler. The safety valve stays open until sufficient steam is released and there is a definite drop in pressure inside of the boiler. This drop in pressure is known as *blowdown* (blowback). Safety valves are made to close without *chattering*. Chattering occurs when a valve opens and closes rapidly.

In most states, the only type of safety valve allowed on steam boilers is the spring-loaded pop type. Many spring-loaded pop safety valves have no adjustments and prevent their normal operation from being changed. See Figure 2-1.

The safety valve on a low pressure boiler opens when 15 pounds of pressure or more are exerted against every square inch of the safety valve that is exposed to the steam. To determine the total force of the steam acting against the safety valve, the following formula is used.

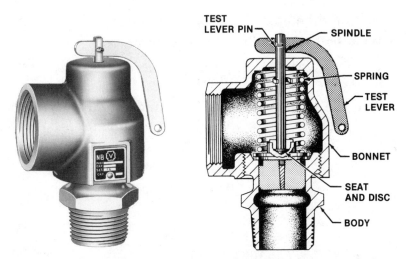

Figure 2-1. The spring-loaded pop safety valve pops open when steam pressure exceeds 15 pounds per square inch in the low pressure boiler. (*Manning, Maxwell, and Moore, Inc.*)

Where *TF* is *total force*, *A* is *area*, and *P* is *pressure:*

$$TF = A \times P$$

For example, if the area of the safety valve is 7 square inches, and if 15 pounds of pressure are pushing against every square inch, then the total force of the steam acting against the safety valve is 105 pounds.

Since the base of the safety valve is a circle, the formula for finding the area of a circle is used to find the area of a safety valve exposed to the steam pressure. That formula is:

$$Area\ of\ a\ circle = diameter^2 \times 0.7854$$
$$A = d^2 \times 0.7854$$

The value 0.7854 is a constant and remains the same no matter what the size of the circle. The *d* in the formula stands for the diameter of the circle and is squared (multiplied by itself).

If the diameter of the safety valve is 3″, the area is determined by applying the formula:

$$A = d^2 \times 0.7854$$
$$A = 3^2 \times 0.7854$$

So the total force that 15 pounds of steam pressure would exert on a safety valve whose diameter is 3″ would be:

$$TF = A \times P$$
$$TF = d^2 \times 0.7854 \times 15 \text{ psi}$$
$$TF = (3^2 \times 0.7854) \times 15 \text{ psi}$$
$$TF = (3 \times 3 \times 0.7854) \times 15 \text{ psi}$$
$$TF = 7.0686 \times 15 \text{ psi}$$
$$TF = 106.029 \text{ lbs.}$$

The spring in the safety valve holds the valve tightly closed against its seat until the steam pressure reaches 15 pounds. The total force of the steam then slowly lifts the valve off its seat. The steam now enters the *huddling chamber.* Because the huddling chamber is wider, this gives the steam a larger area to push against so the total force increases. This relieves the boiler pressure and by opening quickly, prevents the seat of the safety valve from being damaged by steam. See Figure 2-2.

The safety valve must stay open until there is a drop in pressure of usually 2 to 4 psi. The safety valve then closes. The drop in pressure gives the boiler operator a chance to determine why pressure exceeded 15 psi and get the boiler back under control.

The American Society of Mechanical Engineers (ASME) has set up a code that should be followed if the boiler is to function safely. The ASME code specifies the type of material to be used and the location and number of valves, according to the temperatures and pressures at which the boiler operates.

The ASME code states that all boilers with over 500 square feet of heating surface (heating surface is any part of a boiler that has water on one side and gases of com-

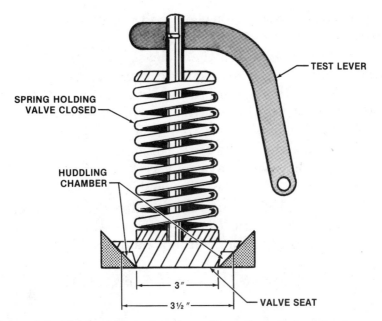

SPRING HOLDING
VALVE CLOSED

HUDDLING
CHAMBER

TEST LEVER

3″

3½ ″

VALVE SEAT

Figure 2-2. The huddling chamber allows the safety valve to open quickly, preventing damage to the valve seat.

bustion on the other side) shall have two or more safety valves. The pressure on the boiler cannot go higher than 5 pounds above the MAWP with all valves popping.

Testing Safety Valves

Safety valves must be tested to assure proper operation. Safety valves are tested by hand by lifting the test lever, or under pressure by bringing the boiler pressure up to the point where the safety valve pops open. When testing the safety valve by hand, there should be at least 5 pounds of pressure on the boiler.

According to Section VI of the ASME code, safety valves on low pressure boilers should be tested by hand once every 30 days the boiler is in operation. With at least 5 psi of pressure on the boiler, the test lever on the safety valve is lifted to the wide open position. The valve is held open for 5 to 10 seconds. See Figure 2-3.

The hand test lever is then released, allowing the valve to snap shut. The ASME code also recommends the testing of the safety valve under pressure once a year, preferably before the start of the heating season. In addition to the ASME code, safety valves should be tested according to recommendations of the local boiler inspector.

Some boiler rooms have devices that aid in the testing of safety valves. Chains or wire cables are attached to the test lever of a safety valve. This allows the operator to stand on the floor and test the safety valve by hand by pulling a chain or cable, without having to climb on top of the boiler.

If the boiler room does have a chain or cable arrange-

Figure 2-3. Safety valves should be tested every 30 days the boiler is in operation according to the ASME code.

ment, it should comply with the ASME code. This code has been adopted in almost every state and states that small chains or wire attached to the levers of pop safety valves and extended over pulleys to other parts of the boiler room may be used, but must be arranged so that the weight of the chain or wire exerts no pull on the lever.

There is no excuse for not testing a safety valve properly. When testing a safety valve, the worst that can happen is that it will leak and have to be replaced. If the safety valve is not tested, it may be faulty without the operator knowing it, which could cause a boiler explosion.

STEAM PRESSURE GAUGE

The steam pressure gauge shows in pounds per square inch (psi) how much pressure is in the boiler. Low pressure boilers are designed to carry a maximum steam pressure of 15 psi. The steam pressure gauge allows the operator to maintain boiler pressure in the safe operating range. The steam pressure gauge must be connected to the highest part of the steam side of the boiler and located to allow easy viewing by the operator.

Steam Pressure Gauge Operation

Inside the steam pressure gauge is an oval tube shaped like a question mark. See Figure 2-4. This tube is called a *Bourdon tube* after the French scientist Eugene Bourdon who helped develop the steam pressure gauge. The open end of the Bourdon tube is connected to the steam side of the boiler. The closed end is attached to connecting linkage that controls the needle position.

As pressure builds up in the Bourdon tube, it straightens out and moves the needle over a scale on the face to indicate the pressure in psi. The range of the steam pressure gauge is the highest pressure above 0 psi shown on the gauge face. The steam pressure gauge range should be 1½ to 2 times the MAWP of the boiler.

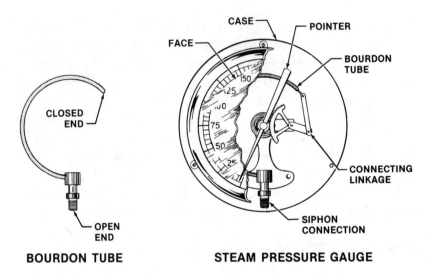

BOURDON TUBE   **STEAM PRESSURE GAUGE**

Figure 2-4. Pressure from the boiler causes the Bourdon tube to straighten in the pressure gauge. Connecting linkage converts movement of the Bourdon tube to the pointer.

The Bourdon tube in the steam pressure gauge is delicate. The steam entering the tube can cause warpage and/or give a false reading. Siphons are installed between the boiler and the steam pressure gauge to prevent damage to the gauge. See Figure 2-5. The two types of steam siphons are the *pigtail siphon* and the *U-tube siphon*.

Both types of steam siphons form a water trap so that water and not steam enters the Bourdon tube. Never blow water out of the siphon and allow live steam to enter the Bourdon tube. Live steam will damage the steam pressure gauge. A gate valve is installed on the pigtail to allow changing the steam pressure gauge in the event of failure or malfunction.

A *compound gauge* is usually used on low pressure boilers. A compound gauge reads *vacuum* in inches of mer-

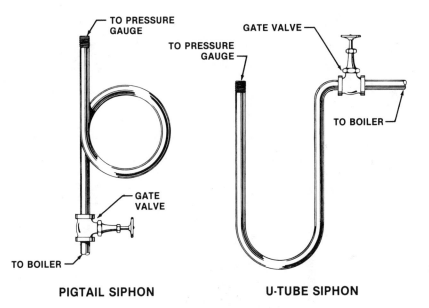

PIGTAIL SIPHON   U-TUBE SIPHON

Figure 2-5. Siphons are used to prevent damage to the Bourdon tube of the steam pressure gauge caused by steam.

cury (Hg) on the left side and pressure in pounds per square inch on the right side. See Figure 2-6.

A vacuum is defined as the pressure below atmospheric pressure. A vacuum gauge is read in inches; for example, a 5″ vacuum or an 8″ vacuum. A pressure gauge is read in pounds per square inch; for example, 5 psi or 8 psi.

Pressure gauges get out of *calibration* through use or misuse, which means they no longer read the proper pressure accurately. A pressure gauge that is out of calibration is either a *fast gauge* or a *slow gauge*. A fast gauge reads more pressure than is actually in the boiler. A slow gauge reads less pressure than is actually in the boiler and is more dangerous than a fast gauge. Any pressure gauge that is out of calibration should be promptly tested. If repairs are necessary, consult the pressure gauge manufacturer.

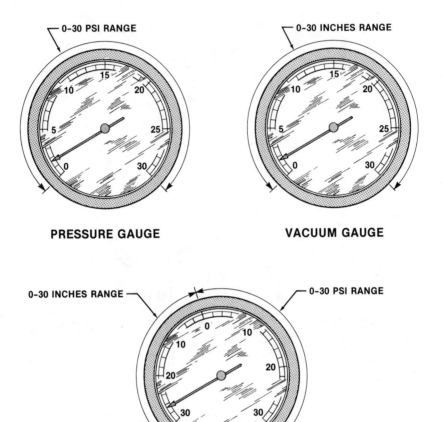

Figure 2-6. Pressure gauges used on low pressure boilers indicate the amount of vacuum or pressure in the boiler system.

Points to Remember

1. The safety valve is the most important valve on the boiler.
2. The safety valve must be connected directly to the shell of the boiler at the highest part of the steam side of the boiler.

3. The safety valve prevents the boiler pressure from going above its MAWP.
4. If the boiler has over 500 square feet of heating surface, two or more safety valves are required.
5. Safety valves should be tested at least once a month to ensure their proper operation.
6. When safety valves pop open, there must be a definite drop in pressure in the boiler before reseating.
7. Safety valves do the same job whether used on a fire tube, water tube, or cast iron sectional boiler.
8. A steam gauge on a boiler indicates the amount of pressure present in the boiler.
9. A siphon protects the mechanism of the steam pressure gauge from live steam.
10. Compound gauges show vacuum on the left side and pressure on the right side of the dial.

WATER COLUMN

A water column is used to reduce the movement of the water so the boiler operator can get an accurate reading of the water level in the gauge glass. The level of the water in the gauge glass indicates the water level in the boiler. When the boiler is steaming, the water inside is constantly in motion, making it difficult to determine how much water is in the boiler. The ASME code does not require that all boilers have a water column, but most steam boilers are equipped with one.

At the bottom of the gauge glass is a *gauge glass blowdown valve.* This allows the operator to blow down the gauge glass lines to remove sludge and sediment and to check the water level. All boilers must have two methods of determining the water level. The gauge glass is the first and easiest method to determine water level.

Try cocks provide the second method of finding the water level in the boiler. See Figure 2-7.

Try cocks are valves that are opened and closed manually. Try cocks can also be weighted valves that are opened by pulling down on a chain. When released the weight on the valve automatically closes the valve.

When opened, try cocks determine the level of water in the boiler. With a normal operating water level (NOWL) of about half a glass, water comes out of the bottom try cock when it is opened. The middle try cock discharges a mixture of steam and water, and the top try cock yields steam. Steam discharged from the bottom try cock indicates a low water level condition in the boiler.

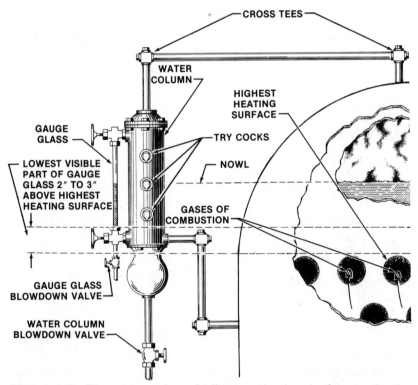

Figure 2-7. The gauge glass indicates the level of water in the boiler. Try cocks on the water column provide a second method of determining water level in the boiler.

The *water column blowdown valve* is used to keep the water column and its lines free from sludge and sediment. Both the gauge glass blowdown valve and the water column blowdown valve should be opened every day.

The location of the water column is very important. The water column must be set at the NOWL so that the lowest visible part of the gauge glass is 2″ to 3″ above the highest heating surface. At this location, as long as water can be seen in the gauge glass, it is safe to add water to the boiler. Never add water to a boiler if water cannot be seen in the gauge glass because a boiler explosion could result.

The boiler operator should be familiar with problems that may occur with the gauge glass, try cocks, and water column. For example, if the top line to the gauge glass is closed or clogged, the gauge glass will fill with water. The gauge glass will read full although the boiler is not full of water. By checking with the try cocks, the boiler operator can determine where the water level is.

If the bottom line to the gauge glass is closed or clogged, the water level in the gauge glass will remain stationary. It will not move up and down as is normal in a steaming boiler. After a period of time the water level will begin to rise in the gauge glass because of the steam condensing on top. This also results in a false water level reading in the gauge glass.

BOTTOM BLOWDOWN VALVES

All boilers, whether they are fire tube, water tube, or cast iron sectional, have bottom blowdown valves on blowdown lines. The bottom blowdown line is located at the lowest point of the water side of a boiler. On a water tube boiler, there may be one or two blowdown valves. When two valves are used, one is usually quick-opening and the other is a screw type. The quick-opening valve is always located closest to the boiler shell. See Figure 2-8.

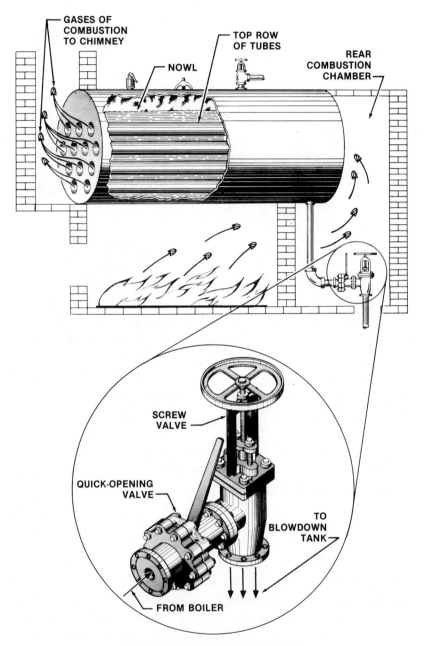

BOTTOM BLOWDOWN VALVES

Figure 2-8. When two blowdown valves are used, the quick-opening valve should always be opened first and closed last.

The quick-opening valve is a lever-operated gate valve that is used as a sealing valve. The screw valve is used as the blowing valve and takes all the wear during blowdown. The screw valve should be constructed so that no pockets of sludge can build up under the seat to prevent it from closing tight. When blowing down a boiler, the quick-opening valve should always be opened first and closed last.

The four reasons for using bottom blowdown valves are

1. to control high water,
2. to remove sludge and sediment,
3. to control chemical concentrations in the water,
4. to dump (drain) a boiler for cleaning or inspection.

Controlling High Water

If the water level in the boiler gets too high, this could lead to *carryover* (water being carried over with the steam into the steam lines). Carryover can cause water hammer and a possible line rupture. To prevent carryover, when there is a high water level in the boiler, the boiler is given a bottom blowdown until it returns to its NOWL.

Removing Sludge and Sediment

Chemicals turn scale-forming salts into a sludge that stays in suspension (suspended in the water). The residue is removed by giving the boiler a bottom blowdown. The best time to blow down a boiler is when the boiler is under a light load.

The water circulation is slower when the boiler is under a light load, and sludge tends to settle to the bottom of the boiler, which makes it easier to remove. The discharge from the bottom blowdown line should not go directly into a city sewer line. The discharge should first go to either a blowdown tank or a sump. See Figure 2-9. This prevents damage to the sewer from the hot water blowdown.

Figure 2-9. The blowdown tank protects the city sewer line from the pressure and high temperature of the boiler blowdown water.

Controlling Chemicals

All water contains certain minerals or scale-forming salts. These salts tend to settle out in the form of scale when the temperature reaches about 150 °F. Scale that has settled on the heating surface of the tubes insulates the tubes so water cannot remove heat from the tubes. As a result the boiler tubes overheat, blister, and eventually burn out.

To prevent scale from forming, chemicals are added to the boiler to change scale-forming salts into a *non-adhering sludge* (sludge that stays in suspension in the water), which can then be removed from the boiler by blowing it down. Chemicals added to the boiler water build up in time, raising the level of concentration. To dilute the concentration of chemicals in the water, the boiler must be blown down and fresh makeup water added.

Dumping the Boiler

Dumping a boiler means to empty it of its contents. Dumping is required because boilers should be examined once a year for defects inside or outside. Before dumping, the boiler has to be taken *off-line* (shut off from the main header) and cooled. After the boiler has been dumped, it should be cleaned thoroughly on both the fire side and the water side.

SURFACE BLOWDOWN LINE

Some boilers have a surface blowdown line. See Figure 2-10. This surface blowdown line is equipped with a valve and located at the NOWL. The surface blowdown line is used to skim off any surface impurities. Surface impurities increase the surface tension of the water. Increased surface tension prevents the steam bubbles from breaking through the surface of the water.

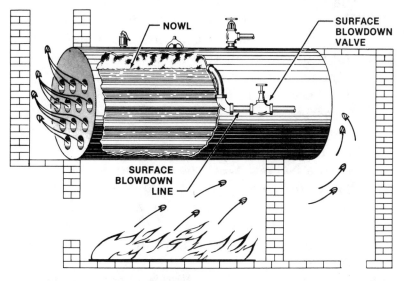

Figure 2-10. The surface blowdown line is used to remove surface impurities on the boiler water.

In addition, increased water surface tension leads to *foaming*, which is a rapid fluctuation in water level. This results in false water level readings. One minute the gauge glass is full of water; the next minute it may be empty. Foaming can in turn also cause *priming* and carryover. Priming is the carrying of small particles of water into the steam lines. By giving the boiler a good surface blowdown, these impurities are removed and foaming is reduced.

Points to Remember

1. The water column is located at the NOWL so the lowest visible part of the gauge glass is 2″ to 3″ above the highest heating surface.
2. The top line of the water column connects to the highest part of the steam side of the boiler. The bottom line connects to the water side well below the NOWL.
3. The gauge glass is located on the water column and provides the easiest method of determining the water level in the boiler.
4. Two or three try cocks are found on the water column and provide a second method of determining the water level in the boiler.
5. If the top line to the gauge glass is closed or clogged, the glass will read full of water.
6. If the bottom line to the gauge glass is closed or clogged, the water level will remain stationary, but fill up slowly.
7. With a NOWL, when opened the bottom try cock has water coming out, the middle try cock has a mixture of water and steam, and the top try cock has only steam.
8. All boilers have blowdown lines equipped with blowdown valves.
9. The bottom blowdown line is always connected to the lowest part of the water side of the boiler.

10. The four reasons for using the bottom blowdown line are (1) to control high water, (2) to remove sludge and sediment, (3) to control the chemical concentration of the water, and (4) to dump or drain the boiler.
11. The bottom blowdown line should first discharge to a blowdown tank or open sump before discharging to the sewer.
12. A surface blowdown line is located at the NOWL and is used to remove surface impurities to prevent foaming.

FUSIBLE PLUG

Although the ASME code now only requires fusible plugs on coal-fired boilers, fusible plugs may still be found on gas and fuel oil fired burners. See Figure 2-11. A fusible plug is the last warning the boiler operator has of low water before the tubes start to burn.

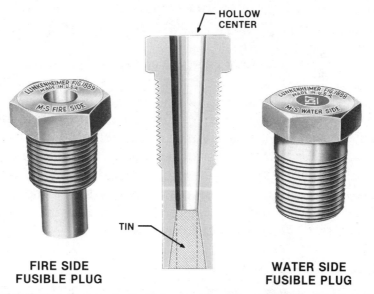

FIRE SIDE FUSIBLE PLUG   **HOLLOW CENTER**   **WATER SIDE FUSIBLE PLUG**

Figure 2-11. The tin in fusible plugs melts at 450°F, warning the operator of a low water level condition. (*Lunkenheimer Co.*)

The fusible plug is a brass or bronze plug with a tapered, hollow center filled with 100% banka tin, which melts at about 450 °F. Fusible plugs are designed for use on either the fire side or the water side of the boiler. The *fire side fusible plug* screws from the fire side into the water side. The *water side fusible plug* screws from the water side into the fire side. The fusible plug is located 1 ″ to 2 ″ above the highest heating surface in the direct path of the gases of combustion.

Fusible Plug Operation

As long as water is in contact with the fusible plug, the tin is prevented from melting. However, if the water level gets low heat is not absorbed, causing the tin in the fusible plug to melt. Steam then escapes through the plug with a whistling noise, warning the operator of the low water level condition.

Both the fire side and water side fusible plugs must be kept clean. Scale allowed to build up on the fusible plug could prevent the water from removing the heat. This may cause the tin in the plug to melt (drop a plug) even though the water level was normal.

Fusible plugs must be replaced annually during boiler inspection. Tin has a tendency to break down with age, changing its melting point. The plug would then be worthless as a warning device. Whenever the boiler is cleaned, remove soot from the fire side of the fusible plug and/or scrape all scale from the water side of the fusible plug.

The fusible plug is generally located in the top of the combustion chamber or in the rear tube sheet above the highest row of tubes in a scotch marine boiler. In a firebox boiler, the plug is in the front tube sheet.

BOILER VENT

A boiler vent (air cock) is used to vent air from a boiler when it is being filled with water. This prevents the air from getting trapped and building up pressure in the boiler. If not properly relieved, pressure from air builds up and will eventually pop the safety valve. The boiler vent also allows air to escape when warming up a boiler before opening the main steam stop valve to the main header. This is called *cutting a boiler in on the line.* The boiler vent also prevents a vacuum from forming in the boiler when taking it out of service.

The boiler operator must clearly understand the danger of a vacuum in a boiler. Steam is a water vapor in a semigaseous condition and occupies space. When a boiler that has been steaming is taken off-line, the main steam stop valve is closed. The steam in the boiler cools and condenses, creating a vacuum. At sea level the atmospheric pressure is 14.7 psi. If the boiler is under a vacuum it is subjected to a force of 14.7 pounds on every square inch of surface.

For example, a simple experiment often used in science classes illustrates the force of this pressure. A 5-gallon can with a small amount of water in it is heated until the water begins to steam. The hole in the can is then sealed and the can is placed under cold water. The steam condenses, forming a vacuum and the atmospheric pressure crushes the can. If a vacuum is allowed to form on a boiler, the boiler is being subjected to unnecessary strains.

If the boiler operator were to try to remove a handhole plate (used for cleaning and washing out the boiler) at the bottom of the water side of the boiler with a vacuum present, the plate could be pulled into the water side. The same could happen with the manhole cover located at the top of the steam side of the boiler. Since the manhole weighs approximately 50 pounds, it could result in damage to the inside of the boiler.

Never attempt to open the steam and water side of a boiler until

1. no steam pressure is on the boiler,

2. the boiler is cool enough to dump,
3. the boiler vent is open and no vacuum is on the boiler.

The boiler vent is a ½ " or ¾ " line with a valve on it coming off the highest part of the steam side of the boiler. Not all boilers are equipped with a boiler vent. If the boiler does not have a boiler vent, the boiler operator must use the try cocks as boiler vents. A safety valve should never be used to vent the boiler because it will result in damage to the safety valve.

Air must be vented from the boiler when warming it up before cutting it in on the line. If the boiler is not vented, air trapped above the water becomes heated, expands, and builds up pressure. The trapped air also affects the pressure reading. The boiler may read adequate steam pressure, although it is air pressure and not steam pressure on the boiler.

The operator, assuming the steam pressure is high enough, may cut the boiler in on the line before it is ready to take its share of the steam load. When the main steam stop valve is opened, the air pressure drops and the steam from the main header rushes in to take its place. Steam pressure then drops throughout the whole system.

PRESSURE CONTROL

The pressure control is a switch that turns the burner on and off based on steam pressure. The pressure control also controls the operating range of the boiler. For example, with an average operating range of 3 to 6 psi, when the steam pressure in the boiler drops to 3 psi, the burner starts up. When the steam pressure reaches 6 psi, the burner shuts off.

The pressure control is located at the highest part of the steam side of the boiler. The pressure control must be protected with a siphon as a steam pressure gauge is.

The pressure control must be mounted in true vertical position to function properly. See Figure 2-12. Some pressure controls have a small, leveling indicator to check for vertical position. Mercury-tube pressure controls are particularly sensitive to correct mounting position. If the

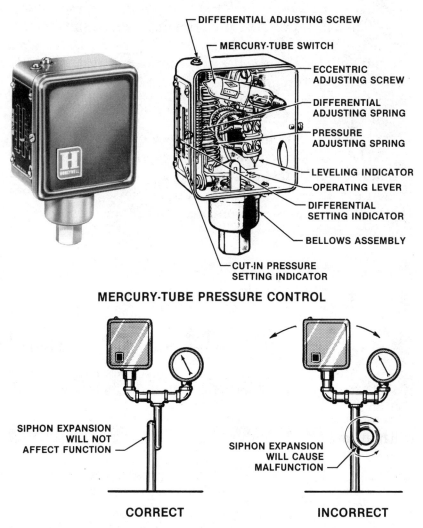

DIFFERENTIAL ADJUSTING SCREW

MERCURY-TUBE SWITCH

ECCENTRIC ADJUSTING SCREW

DIFFERENTIAL ADJUSTING SPRING

PRESSURE ADJUSTING SPRING

LEVELING INDICATOR

OPERATING LEVER

DIFFERENTIAL SETTING INDICATOR

BELLOWS ASSEMBLY

CUT-IN PRESSURE SETTING INDICATOR

MERCURY-TUBE PRESSURE CONTROL

SIPHON EXPANSION WILL NOT AFFECT FUNCTION

SIPHON EXPANSION WILL CAUSE MALFUNCTION

CORRECT   INCORRECT

Figure 2-12. The pressure control turns the burner on and off. It must be installed correctly to function properly. (*Honeywell, Inc.*)

siphon is mounted incorrectly, expansion and contraction of the siphon will cause the pressure control to move. This will result in pressure control malfunction.

Pressure Control Operation

The pressure control has two scales: one for the *cut-in pressure* setting and one for the *differential pressure* setting. The cut-in pressure plus the differential pressure is equal to the *cut-out pressure.* See Figure 2-13. To set a pressure control for the desired operating range, the cut-in and differential pressures are set. For example, for the burner to cut in at 3 pounds and cut out at 6 pounds, the cut-in pressure should be set at 3 and the differential pressure at 3. The burner will then cut out at 6 psi.

NOTE: On some pressure controls the cut-out pressure is determined by setting the main scale set point. The cut-in point is determined by the main scale set point minus the differential setting. Consult the manufacturer for proper setting procedures.

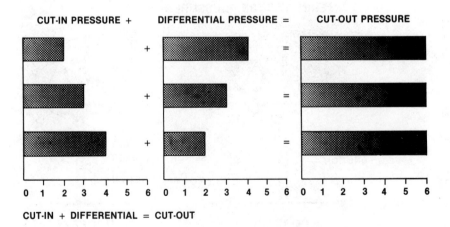

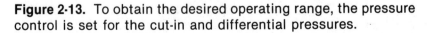

Figure 2-13. To obtain the desired operating range, the pressure control is set for the cut-in and differential pressures.

The mercury tube in the pressure control must be inspected at least twice a year. If the mercury starts to vaporize, conduction of electricity through the mercury will be restricted. As the electricity is restricted the mercury is heated, causing it to raise the heat. As the heat increases, the mercury vaporizes further. Pressure continues to build up, causing the mercury tube to explode. If the mercury tube shows signs of discoloring or the mercury seems to be sticking to the glass in small drops, the mercury tube should be replaced.

The modulating pressure control differs from the pressure control in that it controls both *high fire* and *low fire.* See Figure 2-14. High fire is when the maximum amount of fuel is burned in the burner. Low fire is when the minimum amount of fuel is burned in the burner. A burner should always start off in low fire and shut off in low fire. In addition, the burner should always be firing for longer periods than it is off. This helps to maintain

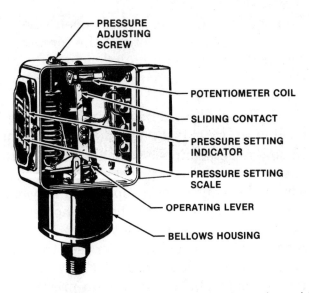

Figure 2-14. The modulating pressure control regulates high and low fire in the burner based on steam demand.

a consistent furnace temperature. Consistent furnace temperature reduces the cooling effect on brickwork, resulting in reduced maintenance and improved boiler efficiency.

Points to Remember _____

1. The fusible plug is usually installed only on boilers burning coal, but it can also be found on boilers burning fuel oil or gas.
2. The fusible plug is installed 1″ to 2″ above the highest heating surface and melts at about 450 °F.
3. The fusible plug is the last warning of low water before a boiler starts burning tubes.
4. ASME code requires fusible plugs on coal-fired boilers.
5. The fusible plug must be kept clean on both fire and water side and replaced annually.
6. The boiler vent is located at the highest part of the steam side of the boiler.
7. The boiler vent must be kept open when filling a boiler with water to relieve the air pressure.
8. The boiler vent must be left open to vent the air from the steam side of the boiler when warming it up.
9. The boiler vent must be open when taking a boiler off-line to prevent a vacuum from forming.
10. In boilers not equipped with a boiler vent the boiler operator must use try cocks as boiler vents.
11. The boiler vent must be open and no vacuum present in the boiler when the manhole cover or the handhole plate is removed.
12. A safety valve should never be used to vent a boiler.
13. A pressure control is an ON/OFF switch that starts and stops the burner on steam pressure.
14. The pressure control must be mounted in true ver-

tical position to function properly.

15. The pressure control is protected with a siphon and must be located at the highest part of the steam side of the boiler.

16. To set the operating range on most pressure controls, the cut-in pressure plus the differential pressure equals the cut-out pressure. (Consult the manufacturer for specific setting procedures.)

17. The mercury tube in a pressure control should be checked at least twice a year for signs of breakdown of the mercury.

18. The modulating pressure control controls high and low fire. High fire is burning the maximum amount of fuel. Low fire is burning the minimum amount of fuel.

19. A burner should always light off (start up) in low fire and shut down in low fire.

20. A burner should always run for longer periods than it is off when burning fuel oil or gas. This helps maintain furnace temperature and reduces cooling of the brickwork.

Key Words

American Society of
  Mechanical Engineers
  (ASME)
blowdown
boiler vent
bottom blowdown valve
Bourdon tube
calibration
carryover
chattering
compound gauge
cut-in pressure
cut-out pressure
differential pressure
fast gauge
fire side fusible plug
foaming
fusible plug
gauge glass
gauge glass
  blowdown valve
high fire
huddling chamber

low fire
maximum allowable
  working pressure
  (MAWP)
mercury tube
modulating pressure
  control
nonadhering sludge
pigtail siphon
pounds per square
  inch (psi)
pressure control
quick-opening valve
safety valve
screw valve
slow gauge
steam pressure gauge
try cocks
U-tube siphon
vacuum
water column
water hammer
water side fusible plug

FEEDWATER ACCESSORIES

3

Feedwater accessories control the quantity, pressure, and temperature of water supplied to the boiler. Maintaining the correct level of water in the boiler is critical. If the water level in the boiler is too high, water can carry over with steam into steam lines.

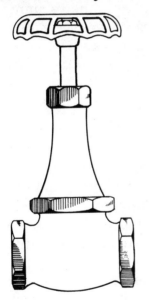

Water in the steam lines can result in water hammer and possible line rupture. If the water level in the boiler is too low, heat from the furnace can burn out boiler tubes and the heating surface. In addition, a low water condition can result in a boiler explosion. For safety and efficiency, the boiler operator must have a thorough understanding of how water is supplied to the boiler.

KEY TO ARROWHEAD SYMBOLS			
△ - Air	▲ - Water	▲ - Fuel Oil	⌂ - Air to Atmosphere
△ - Gas	△ - Steam	△ - Condensate	⌂ - Gases of Combustion

FEEDWATER SYSTEMS

The feedwater system is responsible for supplying water to the boiler at the correct temperature and pressure. In the boiler system, water is converted to steam and converted back to water. See Figure 3-1.

The water in the boiler is heated, turns to steam, and leaves the boiler through the main steam line (1) and goes to the main header (2). Here the main branch line (3) leads from the boiler room and connects with the riser (4) that takes the steam up and distributes it to the various heating units (5).

At this point, the steam gives up its heat to the radiator and turns to condensate (water). The steam trap (6) after each radiator allows the condensate but not the steam to pass through. The condensate flows through the condensate return lines (7) and goes to the vacuum tank (8). The vacuum pump (9) then discharges the water back to the boiler through the feedwater line (10).

Feedwater Valves

Feedwater valves control the flow of feedwater on the feedwater line. The feedwater line has a stop valve and check valve that work together. The *feedwater stop valve* can be opened or closed by the operator to start and stop the flow of feedwater to the boiler. The *feedwater check valve* allows water to flow in one direction, and prevents water from flowing out of the boiler in the feedwater line. See Figure 3-2.

The feedwater stop valve should be as close to the boiler as practical. A check valve should be between the stop valve and the feedwater pump. The valves are arranged so that if the check valve *hangs up* (fails to open or close), it would be possible to close the stop valve and repair the check valve without dumping (draining) the boiler.

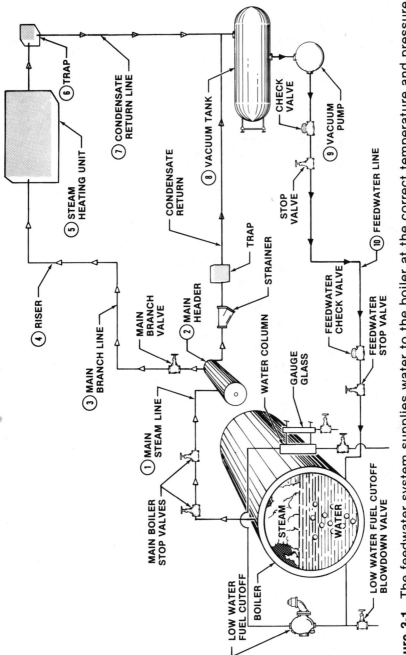

Figure 3-1. The feedwater system supplies water to the boiler at the correct temperature and pressure.

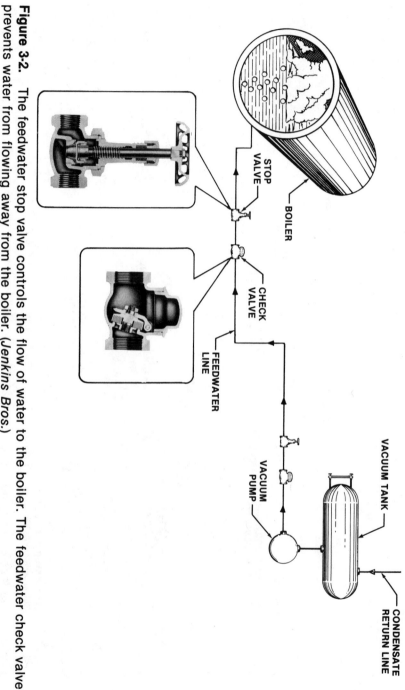

Figure 3-2. The feedwater stop valve controls the flow of water to the boiler. The feedwater check valve prevents water from flowing away from the boiler. (*Jenkins Bros.*)

Feedwater Valve Operation

The feedwater stop valve operates differently on the feedwater line. See Figure 3-3. The feedwater stop valve is opened or closed by the operator by turning the valve handwheel to screw the valve stem in or out. The feed-

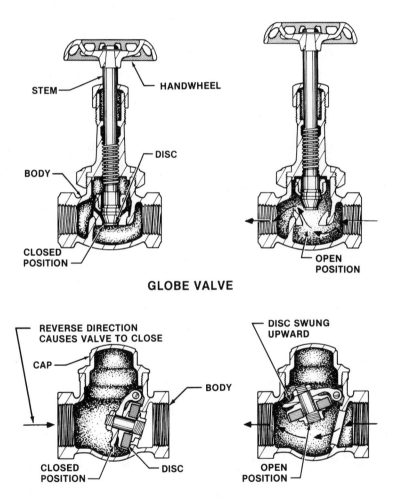

Figure 3-3. The feedwater stop valve is opened or closed by the operator. The feedwater check valve opens and closes automatically.

water stop valve is usually a *globe valve*. A globe valve has a metal disc in the valve that controls the amount of water flowing through it. Although the valve is fully open, the disc impedes the flow of the water passing through the valve. Globe valves are installed when it is desirable to control the rate at which water flows through a valve.

The feedwater check valve is automatic and is not controlled by the operator. Feedwater check valves are usually swing check valves that open and close according to the pressure that acts on them. When feedwater going to the boiler flows into the check valve, the valve disc swings open. Water flowing away from the boiler into the feedwater lines causes the valve disc in the check valve to swing shut.

When the vacuum pump starts and builds up water pressure in the feedwater lines, this pressure overcomes the boiler pressure and the water flows, swinging open the check valve and passing under the seat of the feedwater stop valve before entering the boiler.

When the vacuum pump stops, the boiler pressure becomes greater than the pressure in the feedwater lines. Water then flows away from the boiler and closes the check valve. This prevents boiler water from backing up into the feedwater lines.

Points to Remember

1. To ensure safe operation, the boiler operator must know every possible way of getting water into the boiler.
2. On the feedwater line, the stop valve is located closest to the boiler. The check valve is placed next to the stop valve.
3. The stop valve lets water into the boiler; the check valve allows water to flow in one direction and prevents water from leaving the boiler.

4. The boiler operator can close the stop valve manually if the check valve sticks open or closed. The check valve can then be repaired without dumping the boiler.

VACUUM PUMP

The vacuum pump moves water fom the vacuum tank to the boiler. See Figure 3-4. The vacuum pump serves three purposes:

1. It creates a vacuum on the return lines, drawing back condensate to the vacuum tank.

2. It removes and discharges all the air in the water to the atmosphere.

3. It discharges all the water back to the boiler.

Figure 3-4. Water leaving the vacuum pump is controlled by a stop valve.

Water absorbs some air and a certain amount of this air is released when the water boils or turns into steam. The released air tends to stay in the system, causing rust, corrosion, and pitting of boiler metal. The vacuum pump discharges the air present in water into the atmosphere.

Vacuum Pump Operation

The vacuum pump usually has a selector switch that can be set in three positions: continuous, float only, or float or vacuum. In continuous position, the pump runs continuously. Continuous position is used for testing the pump. In float only position, the pump starts only when the return tank starts to fill up. In float or vacuum position, the pump starts either by a float when the tank become full, or on vacuum when the vacuum falls below a predetermined set pressure.

To test the vacuum pump, the boiler operator throws the selector switch into the continuous position. The vacuum pump is normally in the float or vacuum position during the heating season.

The operation of the vacuum pump is controlled by a vacuum switch similar to a pressure control on the boiler. The range of pressure on the vacuum switch is usually 2 " to 8 ". This means that when the vacuum drops to 2 ", the pump starts whether there is water in the vacuum tank or not. When the vacuum reaches 8 ", the pump shuts off. The vacuum pulls the condensate returns back into the lines.

If the vacuum is *holding* (steady within the operating range) and the condensate is coming back, the tank will start to flood. In this case, the float raises with the rising water level. The pump starts using a float-controlled switch rather than vacuum pressure. The accumulated water is then discharged into the boiler. The vacuum pump returns only the condensate it receives back to the boiler. If any steam is lost due to leaks, the vacuum pump cannot make up the loss of water.

CITY WATER MAKEUP

The vacuum pump can return only the condensate it receives to the boiler. Extra water must be added to the boiler to replace the water that has been lost by blowing down the boiler or through leaks in the system. This extra water is called *makeup water.* Makeup water is added through the city water supply line either manually in the *manual makeup system,* or automatically in the *automatic makeup system.*

Manual Makeup System

Boilers can have both a manual and an automatic city water makeup system or a manual city water makeup system. If a boiler has both, the manual makeup system bypasses the automatic. See Figure 3-5. If the boiler water

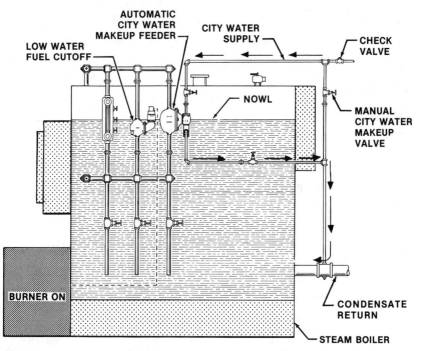

Figure 3-5. In a manual makeup system, the operator opens the manual city water makeup valve to add water to the boiler.

level is low, the operator opens the manual city water makeup valve and city water flows directly into the boiler.

Automatic Makeup System

The *automatic city water makeup feeder* is located slightly below the NOWL. The top line connects to the top of the steam space. The bottom line connects to the water side of the boiler well below the NOWL.

The automatic city water makeup feeder is controlled by a float that is connected to a valve in the city water line. If the water level drops in the boiler, the float drops. This opens the valve in the city water line to feed water into the boiler. As the water level builds up in the boiler, the float rises and mechanically shuts off the automatic city water makeup valve.

Although the automatic city water makeup feeder feeds water to the boiler, it is not meant to act as a feedwater regulator. The function of the automatic city water makeup feeder is to replace water that has been lost. If the automatic city water makeup valve is feeding water to the boiler at regular intervals, the operator must find out why the boiler is not getting the condensate returns back. City water makeup should be added only as necessary.

City water makeup contains scale-forming salts that affect the boiler heating surface unless treated chemically. Chemical treatment is expensive. In addition, the condensate returns coming back are warm and relatively free of oxygen compared to city water makeup, which is cold and contains more oxygen. Excessive use of cold city water makeup reduces overall efficiency because the water must be heated. In addition, oxygen present in untreated city water makeup results in corrosion and pitting of the boiler.

The automatic city water makeup feeder is equipped with a blowdown valve and should be blown down regularly to prevent buildup of sludge and sediment. In addition, a strainer is located on the bottom of the automatic city water makeup valve on the city water line to protect the valve from scale or lime deposits.

This strainer should be cleaned at least once a month, or more often if a rapid buildup of dirt is indicated. The strainer protects the valve and seat from particles of scale or lime deposits that might cause the valve to stick open or closed. If the valve sticks open, the boiler and all the lines would be flooded with water, causing the whole system to become waterlogged. If the valve failed to open, no makeup water could get into the boiler, causing a low water condition.

Points to Remember

1. The vacuum pump pumps water to the boiler and discharges air in the water to the atmosphere.
2. During the heating season, the vacuum pump selector switch is in the float or vacuum position.
3. The vacuum pump returns only the condensate it receives.
4. City water makeup is used to replace water in the system that has been lost due to leaks, blowing down, or failure to receive enough condensate returns back to the boiler.
5. Makeup water can be added manually or by the automatic city water makeup feeder.
6. If the operator finds that the automatic city water makeup feeder is being used too often, the steam and return lines must be checked for leaks.
7. The automatic city water makeup feeder must be blown down at least once a week to prevent buildup of sludge or sediment.
8. The strainer on the city water line before the makeup feeder should be cleaned at least once a month.

LOW WATER FUEL CUTOFF

The low water fuel cutoff shuts off the burner in the event of low water. Although a boiler may be equipped with an automatic city water makeup feeder (not all boilers are so equipped), it is possible that the automatic city water makeup feeder might hang up or city water makeup supply could be interrupted. This could lead to a burned out boiler or a boiler explosion. To protect the boiler, a low water fuel cutoff is used. The ASME code requires low pressure boilers to have a low water fuel cutoff.

The low water fuel cutoff is located slightly below the NOWL. The top line connects to the highest part of the steam side of the boiler. The bottom line connects to the water side, well below the NOWL and is equipped with a blowdown line to keep the float chamber free of sludge and sediment. There are many models of low water fuel cutoffs. As a rule, smaller boilers use small controls and larger boilers use bigger, more intricate controls. The boiler manufacturer specifies the proper type of low water fuel cutoff. See Figure 3-6.

Operation of the Low Water Fuel Cutoff

Despite the many different kinds, the operation of most low water fuel cutoffs is similar. As the water level drops to what would be considered an unsafe level, the float in the low water fuel cutoff drops, and breaks or opens the electric circuit, shutting off the burner. See Figure 3-7.

The burner will fire when the boiler has a NOWL. If the boiler water level drops to an unsafe level, the burner shuts off, preventing damage to the boiler. Note that water is still visible in the gauge glass when the burner shuts off.

Testing the Low Water Fuel Cutoff

The low water fuel cutoff should be tested daily. Testing is done by blowing down the gauge glass, the water col-

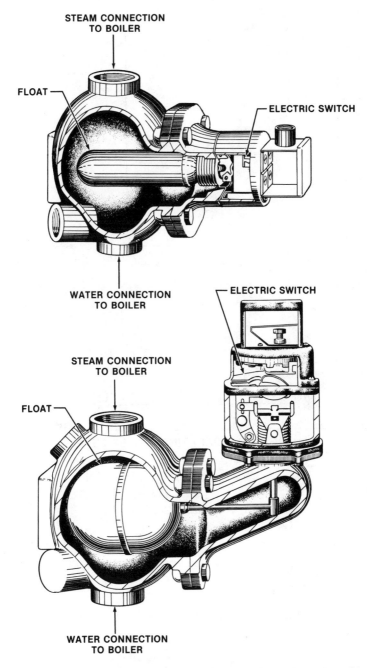

Figure 3-6. The size and type of boiler determines the type of low water fuel cutoff required.

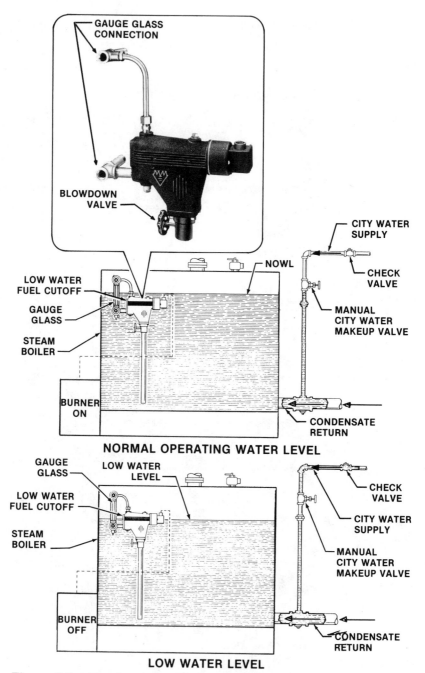

Figure 3-7. The low water fuel cutoff shuts off the burner when water drops to an unsafe level. (*McDonnell & Miller, Inc.*)

umn, and the low water fuel cutoff. The burner should be firing when the low water fuel cutoff is blown down and it should shut off as it is blown down. The boiler operator should always check the water level of all boilers on the line when entering the boiler room.

At least once a month, the low water fuel cutoff should be tested by allowing the water level to drop in the boiler. This is known as an *evaporation test*. This test is done by first shutting off the automatic city water makeup feeder (if the boiler has one) and shutting off the vacuum pump. The burner should be shut off by the low water fuel cutoff at the proper level (about 2") in the gauge glass. The vacuum pump is then started up. When the vacuum tank is empty, the automatic city water makeup feeder should be turned on.

The boiler operator must be present during this test. The boiler operator must carefully monitor the water level in the boiler. If the low water fuel cutoff fails to shut off the burner, an explosion could occur.

Some states have strongly recommended that boilers be equipped with two low water fuel cutoffs. The second low water fuel cutoff should be installed at a level just slightly below the first with separate piping to ensure added protection. A boiler equipped with two low water fuel cutoffs should have both tested as required.

FEEDWATER REGULATOR

Feedwater regulators are accessories that maintain a constant water level in the boiler. In a basic feedwater system, the condensate is returned to a vacuum tank and then continues to the boiler. In a system using a feedwater regulator, after leaving the vacuum tank the condensate flows into a *condensate return tank*, and from there into the boiler. See Figure 3-8.

The condensate (1) returns to the vacuum tank (2). The vacuum pump (3) discharges the condensate to the con-

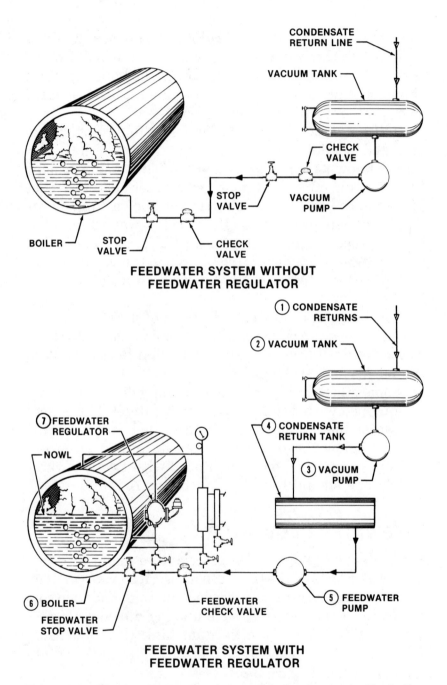

**FEEDWATER SYSTEM WITHOUT
FEEDWATER REGULATOR**

**FEEDWATER SYSTEM WITH
FEEDWATER REGULATOR**

Figure 3-8. The feedwater regulator adds feedwater to the boiler
as required.

densate return tank (4). The feedwater pump (5) sucks water from the condensate return tank and delivers it under pressure to the boiler (6). The feedwater regulator (7) turns the feedwater pump on and off.

The feedwater regulator senses how much water the boiler needs and signals the pump to turn on or off. This regulates the correct flow of water. The feedwater regulator is located at the NOWL and is connected to the boiler in the same manner as the water column and the low water fuel cutoff. The top line is connected to the highest part of the steam side of the boiler and the bottom line to the water side of the boiler below the NOWL.

Feedwater Regulator Operation

When the water level drops in the boiler, the feedwater regulator starts the feedwater pump. When the water in the boiler is at its NOWL, the feedwater regulator stops the pump. The condensate returns are delivered to the return tank by the vacuum pump.

If for some reason the vacuum pump cannot return this condensate and the boiler requires water, the feedwater pump will start. Without condensate in the return tank, the feedwater pump will run dry. This will cause the water level in the boiler to drop. If the low water fuel cutoff works, the burner will shut off because of low water. If the cutoff is not functioning, the boiler could become damaged or explode.

The pump and boiler must be protected from such a low water condition. The condensate return tank protects the feedwater pump and boiler from damage caused by a low water condition. See Figure 3-9. The condensate return tank is equipped with an automatic city water makeup feeder to maintain a constant water level in the return tank at all times.

When the boiler requires water, the feedwater pump draws water from the condensate return tank and is pro-

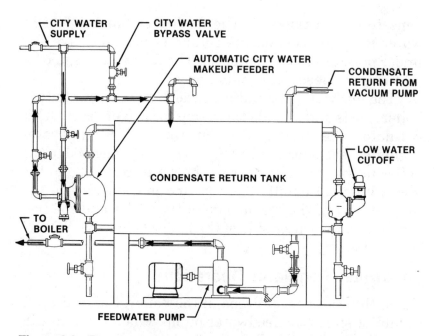

Figure 3-9. The condensate return tank provides a constant source of water to be used in the boiler.

tected from a low water condition. The boiler is also assured of the feedwater it needs. If the automatic city water makeup feeder fails to supply water to the return tank, the low water cutoff will shut off the feedwater pump, preventing it from overheating.

Points to Remember

1. ASME code requires low pressure boilers to have a low water fuel cutoff.
2. Low water fuel cutoffs shut off the burner in the event of a low water conditon.
3. The low water fuel cutoff is located slightly below the NOWL.
4. The type of low water fuel cutoff control required

is determined by the size and type of boiler.

5. Each low water fuel cutoff has a blowdown valve that should be opened daily to test the low water fuel cutoff while the boiler is firing.

6. An evaporation test (dropping the water level in the boiler) should be performed once a month to check the function of the low water fuel cutoff.

7. The boiler operator must be present during an evaporation test.

8. During the evaporation test the automatic city water makeup feeder and the vacuum pump are shut off.

9. If the boiler is equipped with two low water fuel cutoffs, both must be tested for proper operation.

10. In boilers equipped with two low water fuel cutoffs, one must be located slightly below the other with separate piping.

11. The feedwater regulator is located at the NOWL and is connected to the boiler in the same manner as the water column.

12. The feedwater regulator maintains a constant water level in the boiler by starting and stopping the feedwater pump.

13. The condensate return tank collects condensate received from the vacuum tank.

14. The feedwater pumps draws its water from the condensate return tank. Water is supplied to the condensate return tank from the vacuum pump.

15. The condensate return tank has an automatic city water makeup feeder to keep a constant water level in the return tank.

16. The low water cutoff on the condensate return tank protects the feedwater pump by shutting it off if the tank has too little water.

Key Words _____

automatic city water
  makeup feeder
automatic city water
  makeup system
automatic city water
  makeup valve
condensate return tank
evaporation test
feedwater check valve
feedwater pump
feedwater regulator

feedwater stop valve
feedwater valve
globe valve
low water fuel cutoff
makeup water
manual city water
  makeup system
swing check valve
vacuum pump
vacuum switch

STEAM ACCESSORIES

4

Steam accessories are fittings used to control steam in the boiler system. Steam generated in the boiler is piped to various locations to be used for heating or other industrial applications.

Different valves are used to regulate the flow of steam. Gate valves, when open, permit unrestricted flow of steam. Globe valves restrict the flow of steam but are useful in controlling the amount of steam flowing through the line. Steam traps remove air and water from the steam lines without the loss of steam. Air must be removed from steam lines to maintain steam efficiency. Water in the steam lines can result in water hammer or a pipe rupture. Steam traps are tested to ensure proper function. Steam strainers located before the steam traps remove foreign matter from steam that could cause a steam trap malfunction.

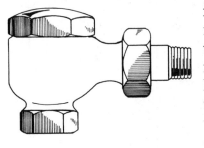

KEY TO ARROWHEAD SYMBOLS			
△ - Air	▲ - Water	▲ - Fuel Oil	∩ - Air to Atmosphere
△ - Gas	△ - Steam	△ - Condensate	∩ - Gases of Combustion

MAIN STEAM STOP VALVE

To cut the boiler in on the line (allow steam to flow from the boiler into the header) or take the boiler off-line, the main steam line must have a main steam stop valve. This valve should be an outside stem and yoke (os&y) gate valve. See Figure 4-1.

An os&y gate valve works just like an overhead door of a garage. To get the car out of the garage, the door is lifted. When the car is back in the garage, the door is pulled down. In an os&y gate valve to let steam pass through it the gate, which is a metal plate, is lifted. To shut off the flow of steam, the gate goes down, closing the valve. When the os&y gate valve is open, the steam flows through the valve and is not restricted.

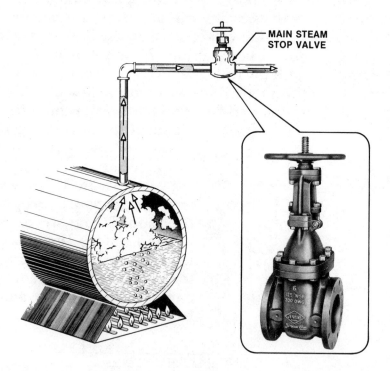

Figure 4-1. The main steam stop valve controls the flow of steam from the boiler. (*Jenkins Bros.*)

A globe valve is never used as a main steam stop valve. See Figure 4-2. When passing through a globe valve, steam flows in under the seat up and out the outlet side. The globe valve restricts the flow of steam as it is routed through the valve. In full open position, the steam still changes direction, causing a drop in steam pressure.

The boiler operator must know exactly when the main steam stop valve and the valves on the header are open or closed. When the valve is 20´or 30´in the air as it often is on large boilers, it is hard to tell whether the valve is open or closed without moving the valve handle.

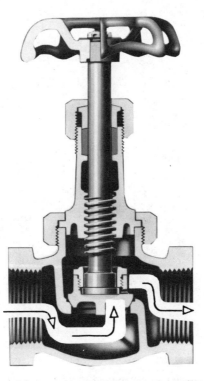

Figure 4-2. Steam changes directions when flowing through a globe valve, causing a drop in pressure. (*Kennedy Valve Co.*)

Os&y valves show the operator at a glance whether the main steam stop valve and the valves on the header are open or closed. See Figure 4-3. When the os&y valve is open, the stem is in the up position. When the os&y valve is closed, the stem is in the down position. Os&y valves are sometimes referred to as rising stem valves.

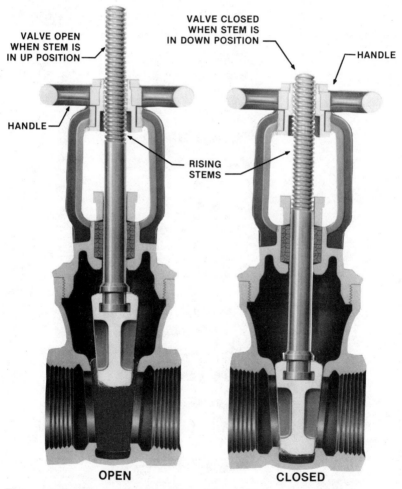

OPEN   CLOSED

Figure 4-3. The position of the rising stem on the os&y valve indicates if the valve is open or closed. *(Kennedy Valve Co.)*

Points to Remember _____

1. All boilers in battery must have a main steam stop valve.
2. Main steam stop valves and the steam header valve should be os&y gate valves.
3. An os&y gate valve shows the boiler operator by the position of its stem whether it is open or closed. When open, the os&y gate valve offers no resistance to the flow of steam.
4. Main steam stop valves and steam header valves are always either wide open or completely closed.

STEAM TRAPS AND STRAINERS

Steam traps are automatic devices that increase the overall efficiency of a plant by removing air and water from the steam lines without the loss of steam. Steam drops slightly in temperature as it travels from the boiler to the main steam header. This causes some condensate to form. The condensate must be removed for a number of reasons. If the condensate is picked up with the steam as it travels through the main steam branch lines, it can result in *water hammer*.

Water hammer is caused by steam pushing ahead of condensate as it travels through the steam lines. When the steam-condensate blend reaches a 90 ° turn, the restriction in the pipe and the sudden change in direction of the mixture results in a loud, hammering sound. Water hammer can lead to a pipe rupture in some cases.

To prevent water hammer, condensate is removed from the steam lines by placing steam traps where condensate buildup could occur. Steam traps should be located on the ends of the main steam header, on the end of the main steam branch line, and on each radiator or heat exchanger, where steam gives up its heat. See Figure 4-4.

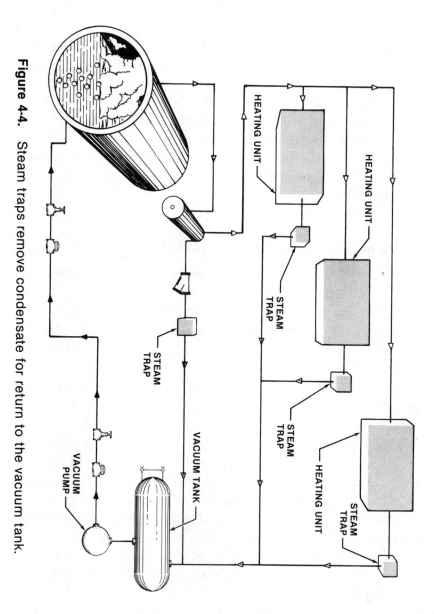

Figure 4-4. Steam traps remove condensate for return to the vacuum tank.

Steam Trap Types

The two types of steam traps are the *return steam trap* and the *nonreturn steam trap*. The return steam trap is no longer used but may occasionally be found on some old systems. The return steam trap is a large trap placed alongside of the boiler and slightly above it. The return steam trap discharges the condensate directly into the boiler when pressure in the trap is equal to or slightly higher than boiler pressure.

The nonreturn steam trap is used on all low pressure steam systems and sends the condensate through the vacuum pump to a condensate return tank. From the condensate return tank the condensate is pumped to the boiler. The three kinds of nonreturn steam traps used are the *inverted bucket steam trap*, the *thermostatic steam trap*, and the *float thermostatic steam trap*. See Figure 4-5.

In the inverted bucket steam trap, steam enters from the bottom and flows into the inverted bucket. The steam holds the bucket up. As the condensate fills the trap up, the bucket loses its buoyancy and sinks, opening the discharge valve. Vacuum from the vacuum tank on the discharge side and the steam pressure from the steam line combine to remove the condensate from the steam trap.

When the condensate has been removed and steam enters the trap again, the bucket is lifted and closes the discharge valve. A small hole at the top of the bucket releases any trapped air. This trapped air would prevent the bucket from sinking and opening the discharge valve.

The thermostatic steam trap is the most common trap and is used on many kinds of radiators. The trap contains a flexible bellows which has a fluid in it that boils at steam temperature. When the fluid boils, the vapors cause the bellows to expand and push the valve closed. When the temperature in the trap falls below steam temperature, the fluid stops boiling and condenses.

The bellows then contracts and pulls the valve into

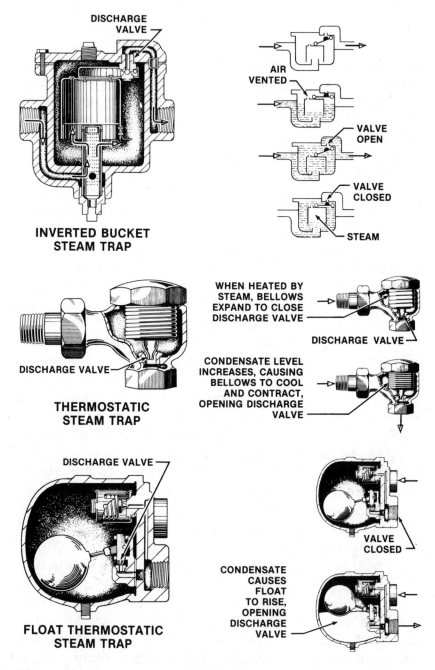

DISCHARGE
VALVE

AIR
VENTED

VALVE
OPEN

VALVE
CLOSED

STEAM

**INVERTED BUCKET
STEAM TRAP**

DISCHARGE VALVE

**THERMOSTATIC
STEAM TRAP**

WHEN HEATED BY
STEAM, BELLOWS
EXPAND TO CLOSE
DISCHARGE VALVE

DISCHARGE VALVE

CONDENSATE LEVEL
INCREASES, CAUSING
BELLOWS TO COOL
AND CONTRACT,
OPENING DISCHARGE
VALVE

DISCHARGE VALVE

**FLOAT THERMOSTATIC
STEAM TRAP**

VALVE
CLOSED

CONDENSATE
CAUSES
FLOAT
TO RISE,
OPENING
DISCHARGE
VALVE

Figure 4-5. Steam traps release condensate through the discharge valve into the condensate return line.

the open position. The bellows expands and contracts, depending on whether it is surrounded by steam or condensate. The opening and closing of the discharge valve allows only condensate to leave the radiator. The vacuum from the vacuum tank and pressure from the steam in the radiator remove the condensate from the trap when the discharge valve is open.

In the float thermostatic steam trap, a float opens and closes the discharge valve according to the amount of condensate in the bowl of the trap. The condensate accumulated in the trap causes the float to rise and open the discharge valve. The vacuum and pressure remove the condensate, causing the float to drop and close the valve. Air in the trap that could interfere with the trap's operation is removed by the thermostatic bellows at the top.

Steam traps must be kept free from dirt and other impurities to function properly. The float and the inverted bucket steam traps have very small discharge orifices. A slight amount of dirt could clog these orifices and block the proper discharge. To prevent this, a *steam strainer* is installed in the line before the trap. See Figure 4-6.

Strainers must be cleaned at regular intervals. Cleaning the strainers is especially important in a new installation because of foreign matter that might be in the pipelines. The strainer should be cleaned every three months for the first year, and twice during the next year. After two years, a thorough once-a-year cleaning should suffice.

Steam Trap Malfunctions

Steam traps require routine maintenance but are neglected more often than any other piece of equipment in the steam heating system. If a steam trap malfunctions and fails to open, condensate will build up. If a steam trap on a radiator becomes waterlogged, the radiator will not heat the room properly.

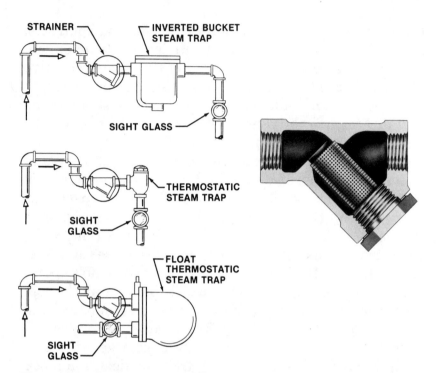

Figure 4-6. Strainers collect foreign matter that could cause the steam trap to malfunction. *(Sarco)*

If the steam trap malfunctions and sticks open, steam will blow through into the vacuum pump or condensate return tank. This can result in the temperature in the room rising because the steam cannnot be properly controlled. Live steam returning to the vacuum tank or condensate return tank raises the temperature of the return water. The condensate pump may become *steambound* (water will flash into steam), which would transform the condensate in the return lines into steam.

This steam cannot be pumped back into the boiler, causing the water level in the boiler to drop and the boiler to shut down because of low water. If the boiler is equipped with an automatic city water makeup feeder, city water will flow into the boiler. After the steambound condition in the condensate pump has been corrected, the boiler operator has to remove the excess water using the bottom blowdown valve.

If steam blows through the malfunctioning steam trap, fuel used to generate steam is wasted. Steam should be present only in the radiators and steam lines. The presence of steam in the condensate return lines reduces boiler efficiency.

Steam Trap Testing

Steam traps should be tested as soon as there is a sign of a room overheating or an increase in temperature of condensate returning to the vacuum tank. Steam traps are easily tested by using *strap-on thermometers* or a *temperature-indicating crayon* (temperature stick). Strap-on thermometers are mounted on the steam line and record the temperature of the line.

A temperature reading should be taken before and after the trap. See Figure 4-7. There should be a 10 °F to 20 °F difference between the two thermometer readings because steam is entering the trap and condensate is leaving the trap. If steam were blowing through the trap, the temperature would read the same on the thermometers before and after the steam trap.

Steam traps in a steam heating system can also be tested using a temperature-indicating crayon. A temperature-indicating crayon has a specific melting point. See Figure 4-8. Temperature-indicating crayons are available in different colors, each having a specific melting point.

A temperature-indicating crayon can be used to test a steam trap by placing a crayon mark on the discharge side of the steam trap. If the steam trap malfunctions and allows steam to pass through, the crayon mark will melt. The specific temperature-indicating crayon used depends on the steam pressure and temperature. As steam pressure

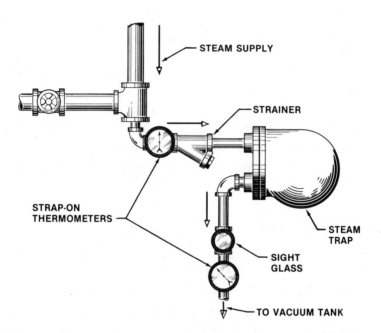

Figure 4-7. Strap-on thermometers should indicate a 10°F to 20°F difference in temperature before and after the steam trap.

Figure 4-8. Temperature-indicating crayons can be used to identify steam trap malfunctions. (*Big Three Industries, Inc., Tempil Division*)

increases, there is a corresponding increase in temperature.

For example, a boiler operating with a steam pressure of 10 psi has a steam temperature of approximately 240 °F. As steam pressure increases to 15 psi, steam temperature is raised to 250 °F. See Figure 4-9. Regardless of the steam

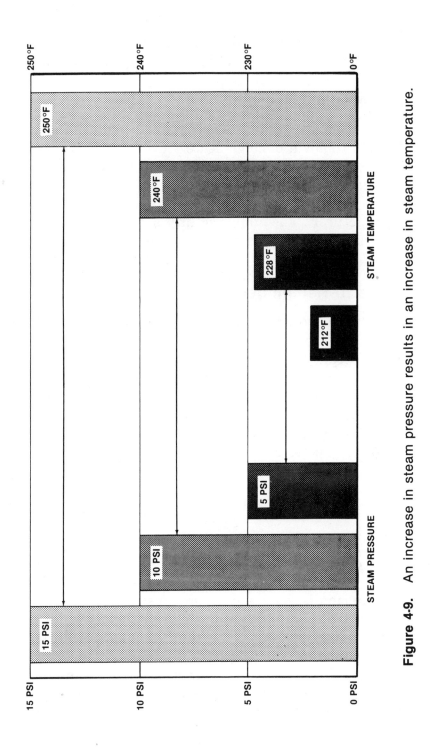

Figure 4-9. An increase in steam pressure results in an increase in steam temperature.

pressure and temperature used, steam traps must be tested to ensure safe and efficient operation of the boiler.

Steam Trap Selection

Steam traps are located as necessary to achieve the most efficient removal of condensate. Several sizes and types of steam traps are available. All steam traps are designed for a specific application in a boiler system. The size and type of steam trap required is best determined by the manufacturer. The manufacturer's field engineer can assist in solving any steam trap problems.

Points to Remember

1. The nonreturn steam trap discharges either to a vacuum pump or to a condensate return tank.
2. A steam trap increases the overall efficiency of a plant and should be tested frequently.
3. Steam strainers should be located in the steam line in front of the steam trap.
4. Steam traps should be tested frequently with strap-on thermometers or a temperature-indicating crayon.

Key Words

float thermostatic
  steam trap
inverted bucket
  steam trap
main steam stop valve
nonreturn steam trap
steambound

steam strainer
strap-on thermometer
temperature-indicating
  crayon
thermostatic steam trap
water hammer

COMBUSTION ACCESSORIES

5

Combustion accessories are used to control the fuel supplied and burned in the burner. In the combustion process, fuel mixed with air is burned to produce the heat necessary to operate the boiler. Fuels commonly used in low pressure boilers include fuel oil, gas, and coal. The type of fuel used is determined by the design of the boiler, the price and availability of the fuel, and compliance with antipollution laws.

Each fuel requires special consideration in storage, handling, and combustion procedures. In addition, the type of fuel used determines the combustion accessories required. All fuels are combustible and can be dangerous if necessary safety precautions are not followed.

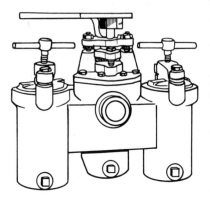

Combustion controls regulate fuel supply, air supply, air to fuel ratio, and removal of gases of combustion.

KEY TO ARROWHEAD SYMBOLS			
△ - Air	▲ - Water	▲ - Fuel Oil	⌂ - Air to Atmosphere
▵ - Gas	△ - Steam	▵ - Condensate	⌂ - Gases of Combustion

FUEL OIL SYSTEM

The fuel oil system supplies fuel oil to the burner at the proper temperature and pressure. Fuel oil must be transported, heated, and regulated before reaching the burner. See Figure 5-1.

Fuel oil is drawn from the fuel oil tank through the suction line (1). A thermometer (2) on the suction line records the temperature of the fuel oil leaving the tank. The duplex strainers (3) remove impurities from the fuel oil. The suction gauge (4) indicates how many inches of vacuum the fuel oil pump is pulling.

The suction valve (5) allows the fuel oil pump to be isolated from the tank for maintenance and repairs. If there is a leak on the suction line between the tank and the suction side of the pump, air will be brought into the system. The suction gauge indicates the leak by pulsating (moving back and forth). The fire in the boiler also pulsates as the supply of fuel oil is interrupted. Air in the suction line must be bled off (removed from the lines).

The fuel oil enters the fuel oil pump (6) and leaves under pressure through the discharge line (7). The relief valve (8) protects the pump in the event of high pressure and discharges fuel oil through the return line (9) to the fuel oil tank. A discharge valve (10) allows the pump to be isolated from the rest of the system for maintenance and repairs.

A pressure gauge (11) indicates the amount of discharge pressure the pump is developing. The fuel oil is pumped to a steam fuel oil heater (13) where the fuel oil temperature is raised. The inlet valve (12) and outlet valve (14) are used to isolate the steam fuel oil heater for cleaning, maintenance, and repairs. A thermometer (15) on the outlet side of the steam fuel oil heater shows the temperature of the fuel oil leaving the heater.

The fuel oil then goes to the electric fuel oil heater (16) where the temperature of the fuel oil is raised to the

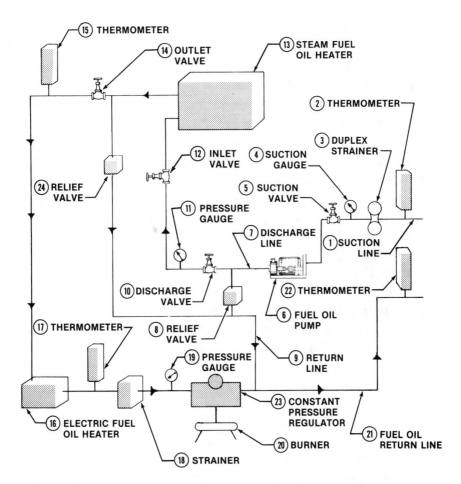

Figure 5-1. Fuel oil must be supplied to the burner at the correct temperature and pressure.

temperature at which it will burn. This temperature varies with the type of fuel oil used. Generally the recommended temperature for burning fuel oil is 150 °F to 180 °F for rotary cup burners and 180 °F to 200 °F for mechanical atomizing burners.

A thermometer (17) shows the temperature of the fuel oil leaving the electric fuel oil heater. Another strainer (18) catches any impurities left in the fuel oil. The pressure gauge (19) located before the burner allows regulation of the fuel oil pressure at the burner (20). The fuel oil is then introduced into the furnace to burn. However, not all the fuel oil is burned. Some of the fuel oil enters the return line (21) and is pumped back to the tank. The thermometer (22) shows the return fuel oil temperature.

The constant pressure regulator (23) maintains the required fuel oil pressure on the system. A relief valve (24) should also be installed on the outlet side of the steam fuel oil heater before the outlet valve discharging back to the fuel oil return line to the fuel oil storage tank. This protects the heater from excessive pressure.

The boiler operator must be familiar with all parts of the fuel oil system. If there is a failure of any part in the system, the boiler operator must be ready to act quickly to remedy the problem. In addition, maintenance and repair procedures necessitate a clear understanding of the flow of fuel oil through the system.

Fuel Oil

Fuel oil consists of carbon, hydrogen, and moisture. In addition, sulfur, nitrogen, arsenic, phosphorous, and silt are present in fuel oil. The different characteristics of fuel oil include *viscosity, flash point, fire point,* and *pour point.*

Viscosity is the internal resistance of the fuel oil to flow. The viscosity of the fuel oil is lowered by raising its temperature. Raising the temperature of fuel oil allows it to be pumped more easily.

Flash point is the temperature at which fuel oil gives off vapor that flashes when exposed to an open flame. Fuel oils with low flash points can be dangerous and require special precautions when handling.

Fire point is the temperature at which fuel oil must be heated to burn continuously when exposed to an open flame. The fire point temperature is higher than the flash point temperature.

Pour point is the lowest temperature at which fuel oil flows as a liquid. If the pour point of the fuel oil is 65 °F and its temperature drops to 64 °F, the fuel oil will start to solidify and cannot be pumped.

Fuel oil must not be overheated in the storage tank. The fuel oil in the storage tank should be between 100 °F and 120 °F. This temperature range allows the fuel oil to be pumped but is not hot enough to give off a vapor that might flash. The temperature of the fuel oil in the tank is controlled by steam supplied to a heating coil or heating bell in the storage tank.

Another consideration in choosing fuel oil is the *heating value.* The heating value of a fuel is expressed in *British thermal units* (Btu). A Btu is the quantity of heat required to raise the temperature of 1 pound of water 1 °F. The Btu rating of a fuel indicates how much heat it can produce.

The American Society for Testing Materials has established standards for grading fuel oils. The four fuel oils most commonly used in low pressure boilers are No. 2 fuel oil, No. 4 fuel oil, No. 5 fuel oil, and No. 6 fuel oil. See Figure 5-2. No. 2 fuel oil is a *distillate* (not mixed with other grades) with a heating value of approximately 141,000 Btu/gal and does not have to be preheated. It is usually burned using a high pressure or low pressure mechanical atomizing burner.

No. 4 fuel oil is heavier than No. 2 fuel oil and has a heating value of approximately 146,000 Btu/gal. Often it is not available as a straight distillate but only as a blend of No. 2 fuel oil and heavier fuel oils. No. 4 fuel oil does not normally require preheating, except in colder climates where it may be necessary to preheat the fuel to lower its viscosity, which facilitates pumping. If the fuel oil is not properly blended, it may stratify (separate) in the tank.

No. 5 fuel oil is divided into hot No. 5, which requires preheating, and cold No. 5, which can be burned straight

Fuel Oil Characteristics	No. 2 Fuel Oil	No. 4 Fuel Oil	No. 5 Fuel Oil	No. 6 Fuel Oil
Type	Distillate	Very Light Residual	Light Residual	Residual
Color	Amber	Black	Black	Black
Lbs/U.S. Gal.	7.206	7.727	7.935	8.212
Btu's/Gal.	141,000	146,000	148,000	150,000

Figure 5-2. Higher numbered grades of fuel oil have a higher Btu content per gallon than lower numbered grades.

from the tank. No. 5 fuel oil has a heating value of approximately 148,000 Btu/gal. In colder climates, tank heaters may be needed to lower the fuel oil viscosity to facilitate pumping.

No. 6 fuel oil, commonly referred to as "bunker C," is a *residual fuel oil*, which means it contains heavy elements from the distillation process. It has a Btu content of approximately 150,000 Btu/gal. Tank heaters and line heaters must be used to heat No. 6 fuel oil to the required temperature for combustion.

Fuel oil has no recommended standard temperatures for burning its various grades. The fuel oil temperature required depends on the type of burners in the plant and whether a straight distillate fuel oil or a blend of fuel oils is used. For best results, obtain the pour point, flash point, and firing point of the fuel oil from the supplier. A rule of thumb is to burn No. 4 fuel oil at 102 °F, No. 5 fuel oil at 150 °F, and No. 6 fuel oil at 220 °F.

Points to Remember _____

1. Fuel oil in the storage tank must be kept below its flash or fire point.
2. The temperature of the fuel oil in the storage tank must be kept above the recommended pour point.
3. Any leaks in the fuel oil line between the tank and the suction side of the pump allow air to get into the system. This causes the fires to fluctuate.
4. The boiler operator should trace all fuel oil lines and become familiar with all valves in order to change over or bypass faulty equipment.
5. A general rule is to burn No. 6 fuel oil between 150 °F to 180 °F in rotary cup burners and between 180 °F and 200 °F in mechanical atomizing burners.

Fuel Oil Accessories

Fuel oil accessories are supporting components that are needed to safely and efficiently operate the fuel oil burner. Fuel oil accessories clean, control the temperature, and regulate the pressure of fuel oil.

Because of varying purity of fuel oil, foreign matter may build up in the lines of the burner, resulting in restricted flow or inefficient combustion. For proper atomization of fuel oil, maintaining the proper fuel oil pressure is essential. The temperature required for efficient combustion varies according to fuel oil grade.

Fuel Oil Heaters. Fuel oil heaters are used to heat certain grades of fuel for ease in pumping. In addition, some grades of fuel oil must be heated before they will burn. To protect the plant, both steam and electric fuel oil heaters should be used.

The temperature of the fuel oil is regulated by the fuel oil temperature regulator. See Figure 5-3. As the temperature increases the sensing bulb (1) picks up or senses the

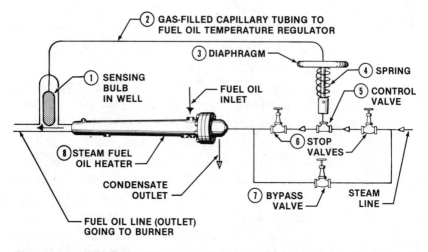

Figure 5-3. The fuel oil temperature regulator controls the fuel oil temperature by the amount of steam passing through the steam fuel oil heater.

temperature increase. The gas in the tube (2) expands and causes pressure on the diaphragm (3).

This pressure overcomes the pressure of the spring (4) and throttles (cuts down) the flow of steam to the steam fuel oil heater through the control valve (5). As the temperature of the fuel oil cools, the sensing bulb senses the temperature drop. The gas in the tube cools, the pressure on the diaphragm drops, and the spring opens the control valve, allowing more steam to enter the steam fuel oil heater (8).

A stop valve (6) on each side of the fuel oil temperature regulator allows repairs. With the fuel oil temperature regulator isolated, the steam to the steam fuel oil heater must be regulated by hand, using the bypass valve (7). The steam fuel oil heater preheats the fuel oil before it reaches the electric fuel oil heater. The electric fuel oil heater located near the burner is used as a booster to raise the fuel oil temperature to the proper burning temperature.

The electric fuel oil heater is also needed when starting up the boiler when there is no steam. The fuel oil pumps are started and fuel oil is circulated through the electric fuel oil heater and back to the tank. When the fuel oil has been warmed enough, the boiler can be lit off (fired) and the steam cycle started.

The temperature of the electric fuel oil heater is controlled by a thermostat. When the proper fuel oil temperature is reached, a switch opens breaking the circuit to the electric fuel oil heater. As the temperature drops, the switch pulls in and closes the circuit to the electric fuel oil heater. A temperature control on top of the electric fuel oil heater can be adjusted to raise or lower the fuel oil temperature.

Fuel Oil Strainers. Fuel oil strainers remove foreign matter in a fuel oil system. Duplex fuel oil strainers should be located on the suction line before the fuel oil pump. Duplex strainers have a fine, mesh screen basket inside. The boiler can be run while cleaning the strainer since only one strainer is cleaned at a time.

The strainers must be kept clean and a clean strainer must always be available. An increase in the suction gauge reading indicates either a dirty strainer or cold fuel oil. The operator should find the problem and correct it. See Figure 5-4.

Fuel oil comes through the inlet (1) into the top of the strainer basket (2) through the sides of the strainer basket and through the plug valve (3), which is tapered. The fuel oil is discharged through the outlet (4). The handle (5) indicates which strainer is in use.

If No. 6 fuel is being used, the strainers should be cleaned every 24 hours. See Figure 5-5. The flange cover (1) is removed and must be cleaned whenever the strainers are cleaned. At this time, the gasket (2) should be replaced carefully to prevent air from getting into the system. The

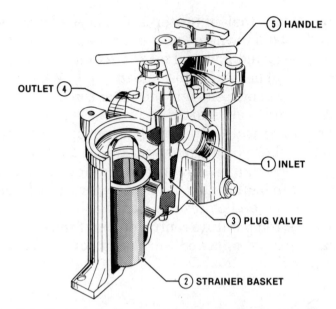

Figure 5-4. Duplex fuel oil strainers allow one strainer to be cleaned while the other is in use. (*The Kraissl Co., Inc.*)

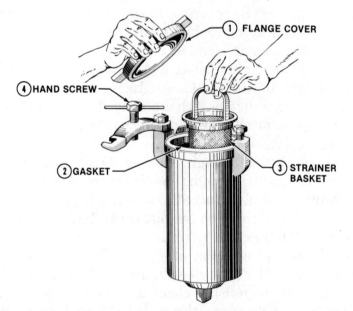

Figure 5-5. Fuel oil strainers must be cleaned more frequently when using heavier grades of fuel oil. (*The Kraissl Co., Inc.*)

strainer baskets (3) have different sizes of mesh screening. When ordering replacement baskets, be sure to order the proper size screening. The flange cover is replaced and tightened down with the hand screw (4).

Fuel Oil Pump.   The fuel oil pump draws the fuel oil from the fuel oil tank and delivers it to the burner at a controlled pressure. Fuel oil pumps may be direct drive or belt drive. See Figure 5-6. For safe and efficient operation of the fuel oil pump, a suction pressure gauge, a discharge pressure gauge, and a relief valve are used.

The relief valve is located between the fuel oil pump and the discharge valve. The suction and discharge valves must be opened before the fuel oil pump is started. The relief valve protects the fuel oil lines and pump if the discharge valve is left closed or if the lines are obstructed.

The relief valve discharges into the fuel oil return line, which returns the fuel oil to the fuel oil tank. A high vacuum of 10 ″ or more on the suction pressure gauge indicates either cold fuel oil or dirty suction strainers.

Fuel Oil Burner.   Fuel oil burners deliver fuel oil to the furnace in a fine spray, which mixes with air to provide efficient combustion. Diffferent types of fuel oil burners utilize various methods to produce the fine spray of fuel oil necessary for combustion.

The types of fuel oil burners commonly used in low pressure boilers today include *rotary cup burners* and *air atomizing burners.* These burner types differ in their accessories, fuel oil pressure, and how they utilize air to function.

The type of fuel oil burned is also a factor in the type of burner used. Rotary cup burners burn No. 4, No. 5, or No. 6 fuel oil at temperatures ranging from 100 °F to 180 °F. Air atomizing burners can burn No. 6 fuel oil at lower temperatures. The fuel oil temperatures identified

DIRECT DRIVE

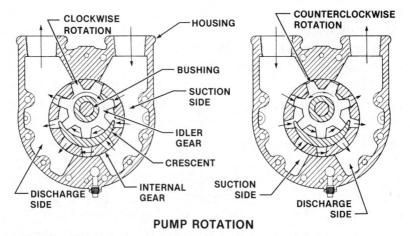

BELT DRIVE

PUMP ROTATION

Figure 5-6. The direction of the rotating gear in the fuel oil pump determines the flow direction of fuel oil. (*The Kraissl Co., Inc.*)

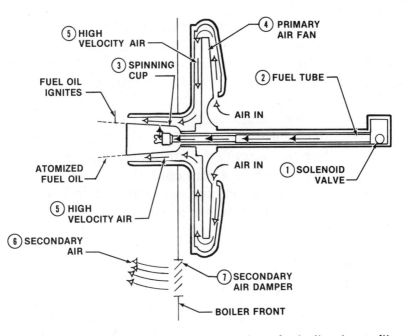

Figure 5-7. The rotary cup burner atomizes fuel oil using a film of fuel oil on a spinning cup.

are approximate. Contact your local supplier to determine the proper temperatures to store and burn fuel oil.

The rotary cup burner atomizes fuel oil using a spinning cup and high velocity air. See Figure 5-7. The solenoid valve (1) controls the flow of fuel oil through the fuel tube (2). Fuel oil flows onto the spinning cup (3), which is rotating, and forms a thin film. The spinning cup increases the velocity of the fuel oil as it moves toward the end of the cup.

The primary air fan (4) produces high velocity air (5) that atomizes the fuel oil as it leaves the spinning cup. Air used to atomize fuel oil is primary air. The secondary air damper (7) regulates secondary air (6), which controls combustion efficiency. The fuel oil pressure required is 5 to 10 psi.

The air atomizing burner can burn fuel oil at a wide range of temperatures. The air atomizing burner uses compressed air to mix with the fuel oil to achieve a high degree of atomization in both light and heavy fuel oil. See Figure 5-8.

The primary (atomizing) air passes around the fuel tube and mixes with the fuel oil in the nozzle. Air and fuel oil in close contact leaving the air atomizing burner allow complete combustion to occur before the flames reach the furnace wall. The secondary air required is kept to a minimum when using an air atomizing burner.

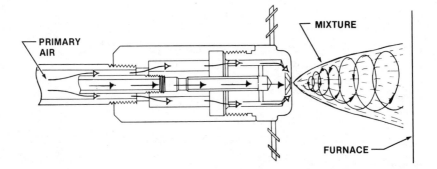

Figure 5-8. The air atomizing burner burns fuel oil at a wide range of temperatures using compressed air. (*Cleaver-Brooks*)

When using No. 5 and No. 6 fuel oil, the burner nozzle and piping are purged so heated fuel oil can flow quickly for the next start-up cycle. Purging of the fuel oil eliminates caking of residual fuel oil in the nozzle and/or solidification of the fuel oil in the piping. See Figure 5-9. Controls are provided to regulate both fuel oil viscosity

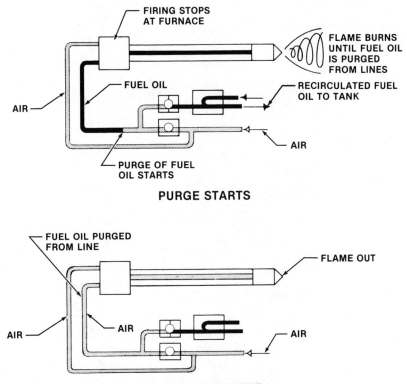

FIRING STOPS
AT FURNACE

FLAME BURNS
UNTIL FUEL OIL
IS PURGED
FROM LINES

FUEL OIL

RECIRCULATED FUEL
OIL TO TANK

AIR

AIR

PURGE OF FUEL
OIL STARTS

PURGE STARTS

FUEL OIL PURGED
FROM LINE

FLAME OUT

AIR

AIR

AIR

PURGE COMPLETE

Figure 5-9. The fuel oil burner nozzle and piping are purged to eliminate clogged lines from solidified residual fuel oil. (*Cleaver-Brooks*)

and pressure to the fuel oil metering valve, thus delivering the exact amount of fuel oil to match the boiler load demand.

The fuel oil burner must be maintained in good operating condition to achieve efficient combustion. Burners foul with deposits of carbon or noncombustible matter that must be removed. See Figure 5-10. To clean the burner nozzle the safety interlock is lifted. The burner is removed and secured in a clamp for cleaning.

LIFT SAFETY INTERLOCK

REMOVE BURNER

SECURE IN CLAMP

Figure 5-10. The fuel oil burner is removed and cleaned as necessary to ensure efficient combustion. (*Cleaver-Brooks*)

GAS SYSTEM

In a gas system, gas burners supply the proper mixture of air and gas to the furnace so complete combustion is achieved. Gas burners in low pressure boilers may be either high pressure or low pressure burners. See Figure 5-11.

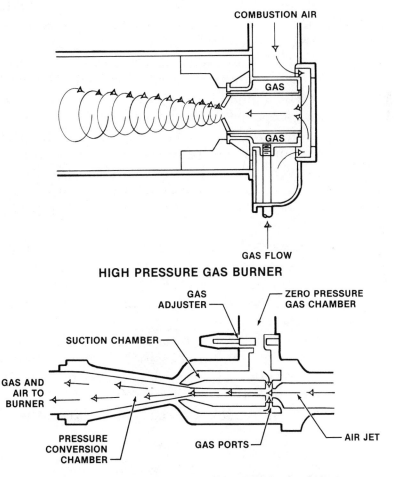

Figure 5-11. In both the high pressure gas burner and the low pressure gas burner, gas is mixed with air for efficient combustion.

In a high pressure gas burner gas is supplied to the burner at a set pressure. The gas mixes with air on the inside of the burner register. In a low pressure gas burner, gas is mixed with air in a mixing chamber before the burner register. The gas is at zero pressure in a low pressure system.

Air passing through the mixing chamber draws gas and provides the air and gas mixture required. See Figure 5-12. The gas line (1) is fitted with a main gas cock (2) that allows the operator to close (cut off) gas to the system when doing any repairs. *NOTE:* The main gas cock is open when the handle is in line with the piping. A solenoid valve (3), which is an electric valve, controls the gas to the pilot (4). The manual reset valve (5) is an electric valve that cannot be opened until the gas pilot is lighted.

The balanced zero-reducing governor (6) reduces the city gas pressure to zero pressure. The small line just before the governor goes to the vaporstat (gas pressure switch) (7). The vaporstat is a switch that is turned on by the gas pressure in the line and turned off when there is no pressure.

The main gas solenoid valve (8) opens at the proper time, allowing gas to be drawn down to the mixjector (9). The blower (10) sends air through the butterfly valve (11). The air passes through the venturi (12) and draws the gas with it into the cage (13).

The block and holder (14) is mounted on the boiler front (15). The air and gas mixture passes through the cage and is ignited by the pilot. The secondary air control ring (16) on the cage controls the secondary air that enters to complete combustion. A pilot gas cock (17) controls the gas flow to the pilot.

Gas Burner Operation

To light off a boiler with a gas burner, the correct sequence must be followed.

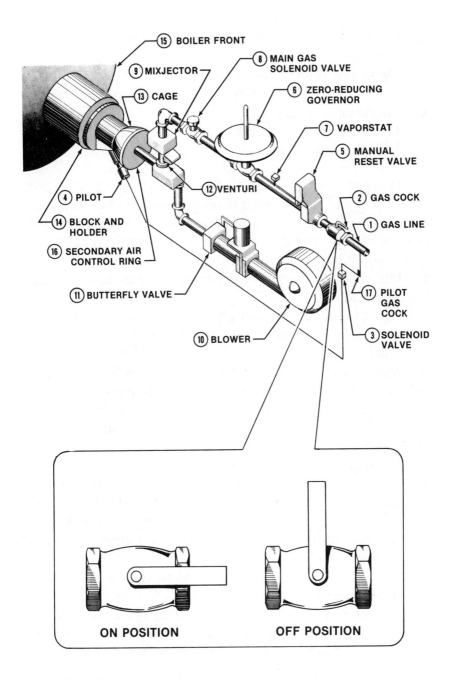

Figure 5-12. The low pressure gas system uses gas reduced to 0 pressure drawn into the mixing chamber.

1. Close the main switch that controls power to all electric controls on the boiler.
2. Open the pilot gas cock.
3. Push the pilot ignition button on the control board and the pilot will light.
4. Open the main gas cock.
5. Lift the handle on the manual reset valve.

The gas now flows up to the zero-reducing governor where it is reduced to zero pressure as it passes through the governor. The vaporstat completes an electric circuit as soon as it proves gas pressure to the governor. The circuit then starts the blower.

The vaporstat proves gas pressure by means of a diaphragm and switch. As the pressure builds up in the gas line, the diaphragm expands causing the mercury tube to tilt, which completes an electric circuit to the blower.

The butterfly valve is a slow-opening valve on the air line that controls the amount of air that can pass through. When the butterfly valve is in the low fire position, only a minimum of air is allowed to flow through. This results in a small amount of air under pressure passing through to the venturi.

As the air passes through the venturi, it pulls in gas and passes the mixture to the cage. Here the mixture is ignited by the pilot and passes through the block and holder into the boiler furnace. Secondary air can be adjusted by opening or closing the secondary air control ring at the front of the cage.

Once the burner has ignited and the boiler is warmed up, the burner may be placed in high fire by putting the selector switch on automatic and pushing the run button on the panel board. The modulating motor will then start to open the butterfly valve, allowing more air to flow through.

The more air that flows, the more gas that is pulled through the mixjector. This puts the boiler in high fire.

The burner is now under the control of the modulating pressure control and the ON/OFF pressure control.

The pilot must ignite before the manual reset valve can be opened to allow gas through. A flame rod must prove the pilot, thus completing an electric circuit and allowing the manual reset valve to engage. In the event of a pilot or main flame failure, the valve closes, shutting off the gas and turning the burner off on safety.

In the event of low water, the manual reset valve also closes, shutting off gas to the burner. In addition, the blower cannot start until the vaporstat located in the line before the zero-reducing governor proves gas pressure in the line up to the governor as another safety precaution.

Points to Remember _____

1. No. 6 fuel oil must be heated to the proper temperature in order to burn.
2. Fuel oil heaters are either steam or electric. Both are needed because the electric must be used to start the boiler before steam is available.
3. Fuel oil strainers must be cleaned and changed every 24 hours.
4. Duplex strainers have two strainer baskets to allow cleaning of one while the other is operating so the flow of fuel oil is not stopped.
5. Fuel oil pumps have suction gauges, pressure gauges, and relief valves. Check these for proper operation of the pump.
6. The atomization of fuel oil in a rotary cup burner is accomplished by the spinning cup and primary air.
7. Air atomizing fuel oil burners mix the fuel oil with air inside the nozzle.

COMBINATION GAS/FUEL OIL BURNER

The combination gas/fuel oil burner can use either gas or fuel oil. Combination gas/fuel oil burners permit the operator to switch from one fuel to another for economy, or because of failure of fuel in use or a shortage of fuel. See Figure 5-13. The combination gas/fuel oil burner has an integral gas burner that allows quick changeover from one fuel to another. Quick changeover is important in an emergency because the boiler can be switched from fuel oil to gas or gas to fuel oil without losing plant load.

Cams control the air to fuel ratio in the combination gas/fuel oil burner. See Figure 5-14. The cams are adjusted for proper air to fuel ratios using adjusting screws on the cams.

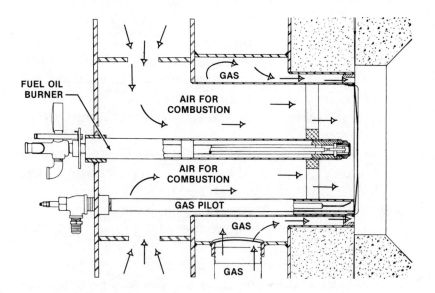

Figure 5-13. The combination gas/fuel oil burner allows the operator to switch from one fuel to another as required.

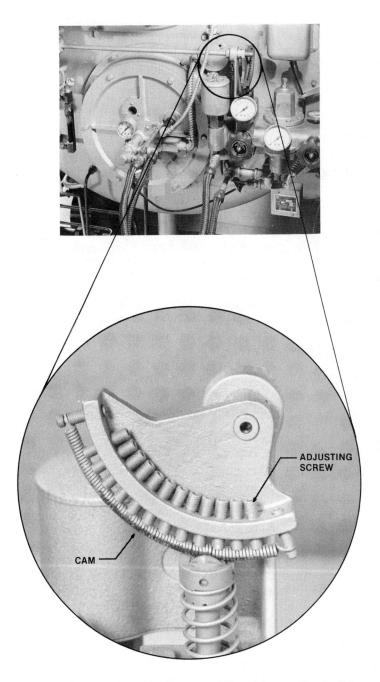

Figure 5-14. Air to fuel ratio in a combination gas/fuel oil burner is controlled by cams. (*Cleaver-Brooks*)

COAL SYSTEMS

Coal systems for steam boilers are used in many large industrial plants. Coal for steam boilers is fed using pulverizers, stokers, or hand firing. Pulverizers use pulverized coal, which is coal ground to the consistency of face powder. Pulverized coal is blown into the boiler where it is burned in suspension as with gas or fuel oil. Pulverizers are commonly used in high pressure boilers. Low pressure boilers using coal use either hand firing or stoker firing.

The type of coal used is determined by the type of furnace. The two types of coal used are hard coal (anthracite) and soft coal (bituminous). Hard coal has a high fixed carbon content and burns on grates used to support the coal bed and allow air to pass through to aid in combustion. Soft coal has a high volatile (gas) content and burns above the grate.

A boiler that uses soft coal as a fuel must have a larger furnace volume than a boiler burning hard coal. Soft coal requires a greater distance above the grates so that the gases can burn completely before they hit the heating surface. Gases that have not burned completely before reaching the heating surface will cool and cause soot (carbon deposits) and smoke. Smoke is a sign of incomplete combustion. Smoke is less of a problem when burning hard coal as compared to soft coal.

Hand Firing Coal

Hand firing is rarely used for feeding coal to the furnace because the size of the boiler must be limited. A boiler operator can feed only a limited amount of coal. When hand firing coal, coal is spread over the grates. A good operator can spread the coal in a fan shape over the surface of the fire. This helps to form an even fuel bed.

When firing by hand, new coal is introduced to the fire

by coaling over. The firing door is opened and the first shovel of coal is thrown on the front of the fire. This serves two purposes. It covers the hot coals and cuts down on the heat that would otherwise blast out at the boiler operator. It also cuts down the intensity of the fire, making it easier for the operator to see into the furnace and evaluate the condition of the fire.

Coal may be fired alternately on one side of the boiler furnace at a time, or it may be fired by coaling over the whole furnace. Firing alternately is preferred because there is less of a tendency to cause smoke because one side of the firebox is always burning to help burn off the volatile gases.

Coaling over often but lightly is also preferred because the operator has better control of the fire as well as the steam pressure. If both sides of the firebox are coaled over at the same time with too much coal, the coal acts as a dampening agent, cooling the fuel bed. This could cause a drop in steam pressure. If a highly volatile coal is used, this would result in smoke.

When soft coal is being hand fired, a shovel, rake, hoe, and slice bar are needed. The Bureau of Mines suggests sizes for these tools. See Figure 5-15. The rake is used for moving coal when firing soft coal. The hoe is used for cleaning the fire. The slice bar is used for breaking up the fuel bed. The fuel bed should be moved and worked when burning soft coal because the burning coal sticks together. This helps the air to get through the grates and aids in combustion. When firing hard coal by hand, the fuel bed must not be disturbed. The fire is kept level with the holes covered up. If the fuel bed is disturbed using hand tools, the fire will drop through holes in the grate.

Hand-fired boilers may have stationary grates, which do not move, or dumping grates to dump the ashes on the grates into the ash pit. When cleaning fires on stationary grates, the boiler operator must pull the ashes and clinkers out either onto the floor or into a wheelbarrow. The fire

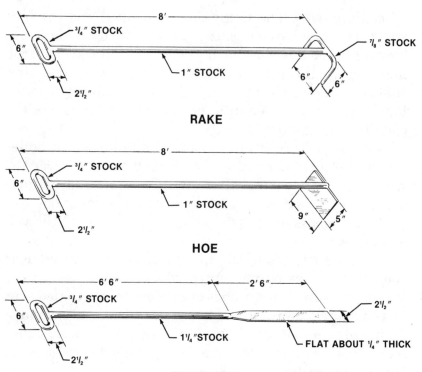

Figure 5-15. Tools are required to work the fuel bed for proper combustion when hand firing bituminous coal.

must be cleaned completely and quickly. If the fire is not cleaned completely, the result is a dirty fire with ashes and clinkers still present. In addition, if the fire is not cleaned quickly, the steam pressure will drop.

Two methods are used in cleaning fires. One method is to pull much of the clean fire to one side. The side to be cleaned should be burned down so as little good coal as possible is lost. Then a hoe is used to pull out all the ash and clinkers. When the grate is completely clean, the clean fire from the other side is pulled over and the firebox is coaled over lightly and the fire built up. The other side is then burned down following the same procedure.

To clean a boiler with dumping grates, the good fire should be pushed to the back, the front grates dumped, and the good fire pulled forward. The fire is built up, the back is burned down, and then the back is dumped. The fire is then spread evenly and built up. Fires must be cleaned as often as necessary. When a fire becomes dirty, it is difficult to maintain the plant load.

Points to Remember

1. Combination gas/fuel oil burners allow flexible operation by permitting the operator to change over to a different fuel as plant needs change.
2. Hand-fired coal should be spread evenly and lightly over the fuel bed. This prevents smoke and slowing down of the fire.
3. Keep fires clean by removing ashes and clinkers as often as necessary. Dirty fires do not heat rapidly.
4. Clean fires in one-half the furnace and then pull the fire over to the clean side and clean the other side.
5. Smoke is a sign of incomplete burning of coal. Too much coal at one time, insufficient air, or improper firing are the likely causes of smoke.
6. Break up the fuel bed when burning soft coal. Use a slice bar or similar tool to open up the coked mass to let air through.

Stokers

Stokers are mechanical coal-feeding devices. The purpose of the stoker is to feed the coal into the furnace of the boiler consistently. Stokers were introduced to increase the efficiency of burning coal by permitting more coal to be burned in the same size boiler. In addition, stokers allowed larger, more efficient coal-burning boilers to be built.

Stokers eliminate the need to open the fire door to coal over. This keeps the furnace at the same hot temperature and chills the brickwork less. Boilers may have one or more stokers, depending on the size of the boiler and the plant load.

Stokers consist of a hopper to hold the coal and a mechanism to feed the coal into the boiler furnace. A fan is necessary to provide air for burning the coal. The most common type of stoker used in low pressure plants is the *underfeed stoker*. The two types of underfeed stokers are the *screw-feed stoker* and the *ram-feed stoker*.

Screw-feed Stoker. The screw-feed stoker uses a short low-speed feed worm screw that conveys the coal from the hopper to the retort. See Figure 5-16. The hopper is designed for easy filling by hand and free flow of coal. Side extensions and a large front apron provide extra large

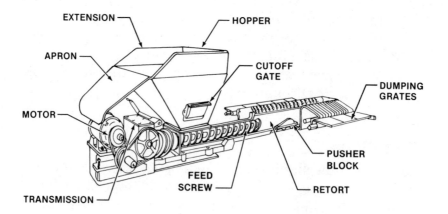

Figure 5-16 A rotating feed screw moves coal from the hopper to the retort in a screw-feed stoker. (*Combustion Engineering Co.*)

capacity. A cutoff gate at the bottom permits removal of foreign material such as stones from the coal-feed mechanism without emptying the hopper.

The feed screw is designed to prevent the coal from packing. It is located entirely outside the fire zone and is free-floating in the conveyor housing. The housing has an access door for inspection or removal of obstructions.

The retort has an expanding mouth that prevents packing of the coal and ensures an easy flow of coal onto the grates. The retort is the full length of the stoker and provides the maximum underfeed combustion area. The retort is located in the main air chamber and is kept cool by the entering air. The retort is anchored only at the front so it is free to expand and contract.

A coal-feed pusher is located in the bottom of the retort and is reciprocating (moving forward and back). It can be adjusted to distribute the coal evenly from front to rear and to provide proper agitation to maintain a porous, level fuel bed that allows air to pass with little resistance.

The grates have an alternating arrangement of moving the stationary bars. The moving bars have a short side motion, which in combination with the reciprocating coal pushers in the retort, provide agitation and even distribution of fuel over the entire grate area. This also provides a gradual movement of the burned out refuse (ash) to the side dumping grates.

Ash dumping grates are located at the furnace side walls at a point furthest away from the entering coal and the hottest combustion zone. The dumping grates are made in sections and are mounted on a heavy bar that is connected to a hand lever at the front of the stoker. These grates are provided with a retaining ledge to hold the fuel bed while the grates are dropped to dump the ashes.

The stoker drive motor operates through a variable-speed transmission that provides for changes in the rate of coal feed. The transmission is enclosed and runs in oil. A shear pin or key in the transmission prevents damage to the stoker in case of an obstruction clogging the feed screw. The shear pin or key is accessible for easy replacement.

The stoker drive mechanism also has a clutch that is used to stop the coal feeder while the fan continues to operate. This permits complete burning down of the fire when banking the boiler or taking it off-line. It also aids in starting a fire by providing air while the coal is ignited.

The forced draft fan supplies air for combustion and is directly connected to the drive motor. The forced draft fan housing is connected to the stoker housing and the volume of air is regulated by the damper on the fan inlet.

Three combustion zones provide for complete combustion over the entire grate area. These zones are the *underfeed zone*, the *moving grate zone*, and the *dumping grate zone*. The underfeed zone at the retort is where the volatile gases are driven off as the coal is heated and gases are burned in the furnace. The moving grate zone is where the coal is burned. The dumping grate zone is where any remaining coal is burned before discharging to the ash pit. Overfire air is provided under separate damper control for complete combustion and smoke prevention.

Ram-feed Stoker. The ram-feed stoker uses a feeder block ram instead of a screw to feed the coal into the retort. See Figure 5-17. The ram-feed stoker does not require as much attention to maintain an adequate fire and efficient combustion over a wide range as the screw-feed stoker. The ram-feed stoker provides uniform feed and even distribution of the coal, quick discharge of ash without disturbing the fuel bed, and proper air distribution and control.

The coal feed ram moves the coal in the hopper into the retort chamber with a reciprocating, sliding bottom. A feeder block or ram on the sliding bottom is mounted at the hopper end and auxiliary pusher blocks are mounted at intervals in the retort chamber.

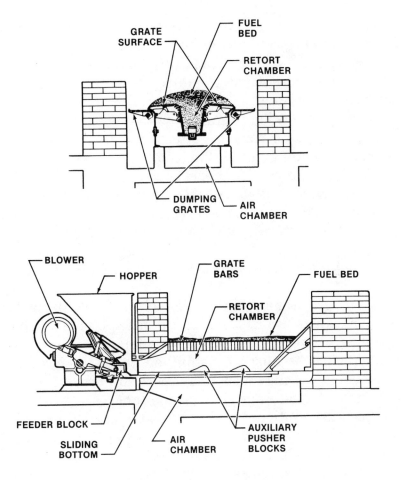

Figure 5-17. Feeder blocks move coal from the hopper to the retort chamber in a ram-feed stoker.

With each upward stroke the feeder block forces a measured quantity of coal into the retort chamber. The auxiliary pusher blocks aid in the distribution of the fuel in the retort chamber and agitate the fuel over the retort.

As the inward stroke progresses, the flow of coal from the hopper is cut off by the feeder block. When the outward stroke occurs, coal falls in front of the block to be pushed into the retort chamber on the next stroke. The amount of coal fed depends on the frequency of the strokes. The length of the stroke remains the same regardless of the rate of coal fed.

Coal distribution is similar to that in the screw-feed stoker. The grate bars are alternately moving and stationary, causing a gradual movement of the coal toward the dumping grates. By properly adjusting the frequency of the stroke of grate bars to the load conditions, the coal is burned before any part of it reaches the dumping grates. A sliding bottom carries the main feeder block and auxiliary pusher blocks that feed coal from the hopper to the rear and at the same time agitates the coal in the retort chamber.

Ram-feed stokers can be driven by a hydraulic piston or gear transmission. The rate coal is fed is controlled by the time between each stroke. The length of the stroke does not vary because a constant length of stroke reduces the tendency of the fuel bed to cake at low operations and provides adequate agitation of the fuel bed.

Coal distribution is maintained at a large throat opening to the retort chamber, full stroke of the feeder and auxiliary pusher blocks, and the sliding bottom. The sideways motion of the grates causes the coal to move gradually toward the dumping grates. The design of the ram-feed stoker provides maximum air pressure over the retort where the fuel is thickest. The air control has three forced draft zones parallel to the retort chamber for control of the air to the grate surfaces, the dumping grates, and over the fire.

Starting a Stoker

When starting up a stoker, the hopper must first be inspected to make sure it is free of any foreign material. Any large stones or metal pieces could cause the shear pin or key to break. A shear pin is provided so the driving mechanism, screw, or ram feed will not be damaged. The stoker should be properly greased and the oil level in the gear box checked. The hopper is then filled and the coal run into the retort. Coal should be fed in until the grates are covered. Wood is laid on top of the coal and waste rags are added.

CAUTION: A volatile liquid should never be poured on top of the wood.

The stoker should be disengaged once enough coal has been put into the boiler. The outlet damper is cracked open and the rags lighted. The forced draft fan should be running with only a small amount of air admitted. Once the coal starts to burn, the stoker feed can be started very slowly and the air flow increased as the boiler is gradually warmed up.

Banking a Fire

When the fire in the boiler is to be banked, the coal feed is disengaged and the fuel bed burned down very low. The forced draft is shut off and the coal runs into the retort until only a small ribbon of fire exists along the whole length of the grates. The outlet damper is cracked open slightly and the forced draft is left off.

Periodically the stoker must be run to introduce green coal into the retort. The fire must never burn down into the retort. This would damage the auxiliary pusher blocks. Some stokers are equipped with an *automatic banking control.* The automatic banking control starts the stoker every half hour and allows it to run from 1 to 10 minutes. This keeps the live fire from damaging the retort and keeps the fuel bed burning.

Burning Down a Fire

Before a boiler is to be taken off-line, the fire must be burned down. A procedure is followed to allow all coal in the hopper to be burned down. After the hopper is empty, it is filled with ashes. The ashes fill the retort, pushing out all the coal onto the grates where it can be burned.

The forced draft is reduced gradually to allow the furnace to cool slowly. As the fire burns down, the water level should be carefully monitored. After the boiler has cooled enough, all the ashes and coal are cleaned from the retort and grates.

Points to Remember

1. Stokers feed coal and air into the boiler where they are burned. The coal feed rate and the air volume are controlled for best combustion.
2. Controls regulate the coal and air. Adjust these according to the fire wanted.
3. To start a fire, fill the retort with coal and then build a fire of wood and rags on top.
4. Never use kerosene, gasoline, or other fluids to start a fire.
5. In shutting down a stoker, be sure to burn all the coal in it and then clean the hopper and retort thoroughly.

COMBUSTION

Combustion is the rapid burning of fuel and oxygen that results in the release of heat. Approximately 14 to 15 pounds of air are needed to burn a pound of fuel. The three types of combustion are *perfect, complete,* and *incomplete.*

Perfect combustion occurs when all the fuel is burned using only the theoretical amount of air. Perfect combustion cannot be achieved in a boiler. It is only possible in a laboratory setting where the combustion process can be carefully controlled. The theoretical amount of air is the amount of air used to achieve perfect combustion in a laboratory.

Complete combustion occurs when all the fuel is burned using the minimum amount of air above the theoretical amount of air needed to burn the fuel. Complete combustion is the boiler operator's goal. If complete combustion is achieved, the fuel is burned at the highest combustion efficiency with minimum pollution. Incomplete combustion occurs when all the fuel is not burned, resulting in the formation of soot and smoke.

Air is necessary for the combustion of fuel. Air contains approximately 20% oxygen and 80% nitrogen. Oxygen will support combustion, but it is not a *combustible.* A combustible will not burn without the introduction of other elements. Nitrogen is not a combustible, nor will it support the combustion process.

Air used in the combustion process is classified into three types: *primary air, secondary air,* and *excess air.* Primary air controls the rate of combustion, thus determining the amount of fuel that can be burned. Secondary air controls combustion efficiency by controlling how completely the fuel is burned. Excess air is air supplied to the boiler that is more than the theoretical amount needed to burn the fuel.

When firing a boiler, the operator's goal is to achieve complete combustion. This involves burning all fuel using the minimum amount of excess air. To obtain complete combustion, the proper mixture of air and fuel, proper atomization, proper temperature of fuel, and enough time to finish the combustion process are required.

This can more easily be remembered by thinking of the letters *MATT. M* stands for mixture of air and fuel. The proper ratio of air and fuel must be maintained at all firing rates. High firing rates burn the maximum amount of fuel and require more air than low firing rates. *A* represents the atomization of fuel. Atomization is the breaking up of fuel into small particles so that it can come into closer contact with the air to improve combustion.

T stands for temperature. The proper temperature of

the air, fuel, and the furnace must be maintained to achieve complete combustion. The second *T* stands for the time needed to achieve complete combustion. The combustion process must be completed before the gases of combustion come in contact with the heating surface.

The boiler operator must maintain efficient combustion to minimize the amount of smoke produced. Efficient combustion reduces fuel costs and air pollution. If combustion is not completed before gases come in contact with the heating surface, the gases will cool and soot and smoke will result.

In addition, soot buildup on heating surfaces acts as insulation and prevents the transfer of heat to the water. This results in an increase in temperature of the gases of combustion discharged to the chimney.

AUTOMATIC COMBUSTION CONTROLS

The purpose of automatic combustion controls is to help maintain a proper air to fuel mixture and control the *firing rate* of the fuel. The firing rate is the amount of fuel the burner is capable of burning in a given unit of time. *High fire* is burning the maximum amount of fuel in a given unit of time. *Low fire* is burning the minimum amount of fuel in a given unit of time.

The automatic combustion control most commonly used on low pressure heating boilers is the *ON/OFF combustion control.* The ON/OFF combustion control regulates the burner by the amount of steam pressure in the boiler. When the steam pressure drops to a certain preset pressure the burner starts, and when the steam pressure reaches a preset pressure the burner shuts off.

Combustion controls regulate

1. fuel supply in proportion to steam demand,
2. air supply,

3. ratio of air to the fuel supplied.

The fuel supply is regulated in proportion to steam demand. For example, on a cold day more steam is required to heat a building. In order to produce more steam, the burner must burn more fuel by maintaining a high fire. As the demand for steam decreases the boiler produces less steam with the burner reduced to low fire.

The ON/OFF combustion control regulates air supply. Both primary and secondary air are needed to burn fuel. Primary air controls the amount of fuel oil capable of being burned. Secondary air controls the combustion efficiency and is usually introduced into the furnace from below the burner.

The air to fuel ratio must be regulated because more fuel burned will require more primary and secondary air. As steam demand increases, the burner will go into high fire. The fuel oil valve is then opened through linkages, allowing more fuel oil to flow to the burner. At the same time, a damper must be opened allowing more primary air to flow to the burner and mix with the fuel oil.

The linkage also opens the secondary air damper, introducing more air to complete combustion. As the load decreases the fuel oil valve closes, throttling (reducing) the flow of fuel oil to the burner. Primary and secondary air are also reduced as the fuel supplied is reduced.

Burners using No. 4, No. 5, or No. 6 fuel oil use a gas pilot to light the fuel oil. The pilot is ignited by a spark plug that is energized by an ignition transformer. Boilers burning No. 2 fuel oil use electrodes that are energized by an ignition transformer to light the fuel oil. No. 2 fuel oil is expensive and is used mainly in homes. No. 4, No. 5, and No. 6 fuel oils are frequently used in larger plants.

The spark from the pilot ignitor lights the gas pilot, which lights the fuel oil. The fuel oil valve is then opened allowing fuel oil to flow through the fuel tube assembly, into the burner and the combustion chamber.

Programmer. The programmer is the mastermind that controls the firing cycle. The programmer has a shaft with fiber cams. See Figure 5-18. As the shaft rotates, phosphor bronze leads containing silver contacts ride on the cams and make electrical contacts at the proper time. These contacts energize circuits that control the operation of the burner in the proper sequence.

When the operating range is 3 to 8 psi, the modulating pressure control is set at 5 psi. When the steam pressure drops to 3 psi, the pressure control completes an electrical circuit to the programmer, which starts to turn. The first contact on the first cam completes a circuit, which starts up the burner motor that rotates the fan and rotary cup.

The fan blows into the furnace and purges any gas or fuel oil vapors, which takes about 30 seconds but may take as long as 60 seconds depending on the type of burner and the programmer. Purging a furnace means to remove any unburned fuel in a gaseous condition in the furnace that might ignite and cause a furnace explosion. Because the furnace is hot, any unburned fuel in the furnace would vaporize because of the heat.

Fuel that is vaporized can ignite if exposed to an open flame. If vaporized fuel accumulates and ignites, an explosion in the furnace would occur. A furnace explosion is just as destructive as a boiler explosion.

After purging, the programmer is still turning and the second contact closes, completing a circuit to the ignition transformer that causes a spark in front of the gas tube. At the same time it opens a solenoid valve in the gas line, allowing gas to flow through the tube and be ignited by the spark. The gas flame burns in front of the rotary cup.

The programmer continues to rotate and the next contact completes a circuit that opens the fuel oil solenoid valve, allowing fuel oil to be pumped into the furnace. The fuel oil is ignited by the gas flame. The programmer is still rotating and the cam that closed the circuit to the gas and spark is opened. The circuit is broken and the sol-

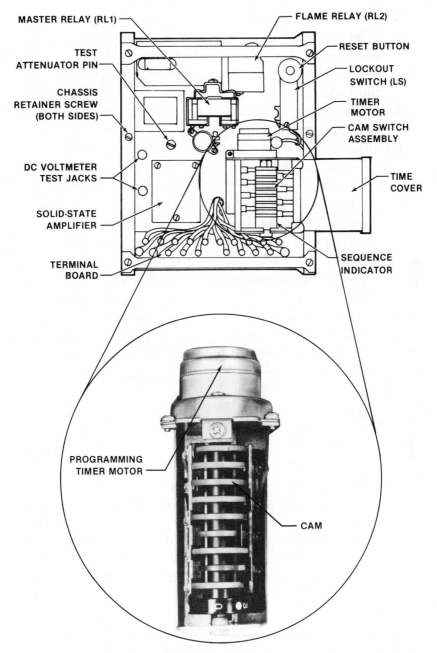

Figure 5-18. Cams inside the programmer activate electrical circuits to control the sequence of operation of the burner. (*Electronic Control*)

enoid gas valve is de-energized, causing the gas valve to close. The spark stops and the burner is self-sustaining.

The programmer completes its cycle and stops. The burner is now under the control of the pressure control and modulating pressure control. If the burner starts to go into high fire, the steam pressure begins to rise. When the pressure reaches 5 psi, the modulating pressure control causes the burner to go into a lower fire.

If the steam pressure in the boiler continues to rise, the burner goes into its lowest fire position until the pressure reaches the shutoff point, which in this case is 8 psi. The pressure control then shuts off the burner in low fire. The programmer is started up again and the fan purges fuel vapors left in the furnace. This is called a postpurge. The programmer stops after the postpurge.

The boiler operator must understand the firing cycle of the burner and know how long it takes the gas pilot to light and the fuel oil valve to open. Understanding the firing cycle allows the operator to monitor the combustion control for proper operation.

Microcomputer Burner Control System.   The microcomputer burner control system (MBCS) performs all of the functions of a conventional programmer. See Figure 5-19. In addition, the MBCS provides improved combustion efficiency, system self-diagnosis, and increased energy conservation.

Combustion safety is the primary function of the MBCS. The control of high fire purge, supervised low fire start, and tamper resistance are performed as necessary. In addition, a mandatory postpurge is initiated if lockout occurs or the fuel valves are energized.

System self-diagnosis functions of the MBCS include reporting of a safety shutdown, elapsed time during prepurge, ignition trials, postpurge sequences, and faults in the flame detection system. Lights illuminated on the

Figure 5-19. A light on the microcomputer burner control system programmer panel indicates the sequence status. (*Honeywell, Inc.*)

MBCS indicate the sequence status: standby, prepurge, hold, ignition trial, flame on, run, postpurge, and safety shutdown.

The energy conservation functions of the MBCS include an energy saving prepurge and energy saving intelligence. The energy saving prepurge prevents blower operation at start-up until the damper reaches purge position (high fire purge switch closed). The energy saving intelligence terminates the burner/blower operation and energizes the alarm circuit whenever the high fire purge switch, low fire start switch, or the running interlocks fail to close after a sufficient time delay. The MBCS can be used on automatically fired fuel oil, gas, coal, or combination gas/fuel oil single-burner applications.

Flame Scanner. The flame scanner (fireye) is a type of control that functions as a safety device in preventing furnace explosions. The flame scanner verifies that the pilot and the main flame are lit. The most common type of flame scanner uses a small lead sulfide cell, which is sensitive to the infrared rays of a flame and is positioned to sense the pilot light and main burner flame. See Figure 5-20. The correct alignment of the flame scanner is required for proper function.

When an increase in steam pressure is required, the pressure control activates the programmer to start the purge cycle and lights the gas pilot. The flame scanner proves pilot (senses the pilot light) and allows the cycle to continue. The fuel oil valve opens, the fuel oil ignites, the flame scanner proves main flame and holds the circuit in, and the programmer continues carrying out its prescribed functions. See Figure 5-21.

If the scanner does not prove pilot during the pilot proving period, the pilot solenoid valve will be de-energized. This breaks the circuit and the solenoid (magnetic) fuel oil valve will not open, thus preventing fuel oil from being pumped into the furnace without a source of ignition.

If the main flame fails, the scanner also shuts off the solenoid fuel oil valve, putting the burner into a flame

Figure 5-20. A lead sulfide cell used in the flame scanner senses infrared light to prove the pilot and main flame. (*Electronic Control*)

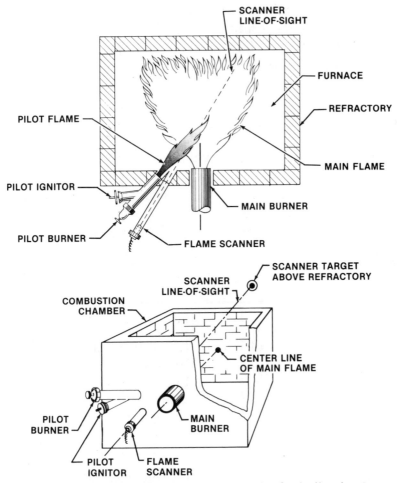

Figure 5-21. The flame scanner controls the fuel oil valve to prevent fuel oil from being pumped into the furnace without a source of ignition.

failure condition. This also prevents a buildup of fuel oil in the furnace. The only way the cycle can be started again is by pushing a reset button in the programmer.

The boiler operator should be familiar with the operation of the flame scanner and the programmer in the firing cycle. Consult the manufacturer of the programmer for complete information on its equipment. Trouble-shooting information provided by the manufacturer will detail where to look for failures and how to correct them. In addition, replacement and testing information is also available.

Points to Remember _____

1. A burner should always light off in low fire and shut off in low fire.
2. A burner must be purged of all fumes or gas after each firing cycle.
3. The flame scanner proves pilot and main flame.

Key Words _____

air atomizing burner
British thermal unit (Btu)
coaling over
coal fuel system
combination burner
combustion
complete combustion
duplex strainer
electric fuel oil heater
excess air
fire point
flame scanner
flash point
fuel oil burner
fuel oil grade
fuel oil heater
fuel oil pump
fuel oil system
gas burner
hand firing

hard coal
heating value
high fire
low fire
MATT
ON/OFF combustion
   control
perfect combustion
postpurge
pour point
primary air
programmer
purge
ram-feed stoker
rotary cup burner
screw-feed stoker
secondary air
soft coal
steam fuel oil heater
underfeed stoker

DRAFT CONTROL · 6

Draft is required for the necessary flow of gases to and from the furnace. For the boiler to operate, there must be a flow of air to the fuel, a flow of hot gases of combustion to the heating surfaces, and a flow of gases of combustion up the chimney. Air at the correct temperature and pressure combines with the fuel and burns to generate heat. Heat in the gases of combustion contacts the heating surfaces, causing the boiler water to boil into steam. The gases of combustion must be removed to make room for incoming gases. The correct draft will maintain the flow of gases to and from the furnace for efficient combustion.

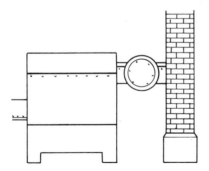

The amount of draft determines the rate of combustion. Too much draft wastes fuel by sending heat up the chimney, while not enough draft results in soot and smoke.

KEY TO ARROWHEAD SYMBOLS			
△ - Air	▲ - Water	▲ - Fuel Oil	⌂ - Air to Atmosphere
⚠ - Gas	△ - Steam	⚠ - Condensate	⚠ - Gases of Combustion

DRAFT

Draft is the difference in pressure between two points that causes air or gases to flow. To have a flow there must be a difference in pressure between two locations. Air or gases flow from the location with higher pressure to the location with lower pressure.

The two types of draft that can be produced in a boiler are *natural draft* and *mechanical draft*. Natural draft is produced by the natural action of heated air or gases. Mechanical draft is produced by a fan or blower that creates a draft. Natural draft is not used with automatic fuel systems because of the amount of draft required. In addition, natural draft cannot be regulated as precisely when compared to mechanical draft.

Natural Draft

Natural draft applies one of the basic laws of physics: warm air rises. As it is warmed, air expands and becomes lighter. Colder, heavier air pushes in under it and forces the warmer air up. The movement of warm and cold air creates draft. By directing the flow of air or gases into a pipe, the amount of draft is increased or decreased. Fireplaces, wood-burning stoves, and some smaller boilers use natural draft to supply the necessary *air for combustion.* Air for combustion is air provided to the furnace to promote the combustion process.

Natural draft is regulated by controlling the air for combustion and heated gases of combustion. The more a gas is heated, the faster it rises. A column of heated gas weighs less than a column of cool air. Thus the higher the column of gas, the more speed it develops as the weight of the column of cold air increases.

On a chimney that is 10 ′ high, when paper is burned in its bottom to heat the air in the chimney, the air rises and cold air rushes in at the bottom to take its place. The amount of natural draft generated is determined by the height of the chimney. See Figure 6-1.

If the height of the chimney is increased, the column of warm air or gases of combustion is also increased and more air will rush in the bottom to take its place, creating a stronger draft. If the chimney is closed at the bottom and the only way for the cold air to enter is through a pipe,

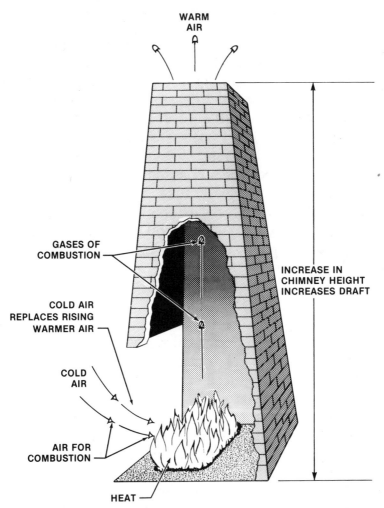

Figure 6-1. In natural draft, hot air and gases of combustion flow upward and are replaced by cold air for combustion.

the air going to the fire can be controlled. Air can then be routed as necessary through the boiler and breeching.

If air comes in contact with the burning fuel, a hotter fire is made by providing more air for the fuel to use. If air is allowed to enter the chimney without contacting the burning fuel, the amount of air the fuel is receiving can be reduced, which will keep the fire smaller. Natural draft is regulated by the amount of heated gases of combustion going up the chimney and by the height of the chimney.

By adding cold air to the gases going up the chimney, the heat of the gases is reduced and their movement is slower. Dampers or draft regulators placed after the fire let air pass directly into the chimney to dilute the heated gases, slowing the heated gases down and reducing the amount of air being pulled into the fuel. If the dampers are closed, the air has to pass through the fuel, increasing the rate of combustion. An increased combustion rate causes the fire to become hotter and the gases to rise faster, creating more draft. See Figure 6-2.

Natural draft was used extensively on hand-fired coal boilers and on some older fuel oil fired boilers. Modern boilers are designed to use mechanical draft, which results in higher combustion rates and more combustion efficiency.

Mechanical Draft

Mechanical draft is produced by power-driven fans. The two types of mechanical draft are *forced draft* and *induced draft.* Forced and induced draft can be compared to how a vacuum cleaner works. Most vacuum cleaners have two places to attach the hose. Normally the hose is attached to the vacuum cleaner so the fan and motor are in back of the hose. A vacuum is created as air is pulled through the hose to the fan and motor.

Induced draft is air that is pulled through the boiler into the induced draft fan. If the hose is attached to the other opening of the vacuum cleaner, it will be after the

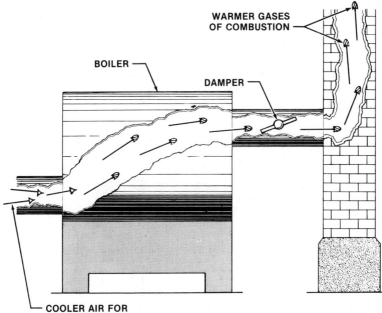

WARMER GASES
OF COMBUSTION

BOILER

DAMPER

COOLER AIR FOR
COMBUSTION

Figure 6-2. Draft in a natural draft system can be controlled by a damper in the breeching.

fan and motor, and the air will be forced out of it. This is how forced draft works in a boiler.

Forced draft is produced when air is forced into the burner with a fan located in front of the boiler. If the air is forced through the fuel in a boiler, the fire will burn hotter because more air will be available for combustion. Induced draft is produced by putting a fan in the breeching or smoke pipe after the burner so the air is pulled out of the boiler and pushed up the chimney. This creates vacuum (suction) in the furnace, causing air to take the place of the air that has been forced out the chimney. See Figure 6-3.

When using natural draft, the boiler is limited in the amount of fuel that can be burned because there is no method to accurately control the draft. In winter, a greater draft is possible because of the difference in temperature

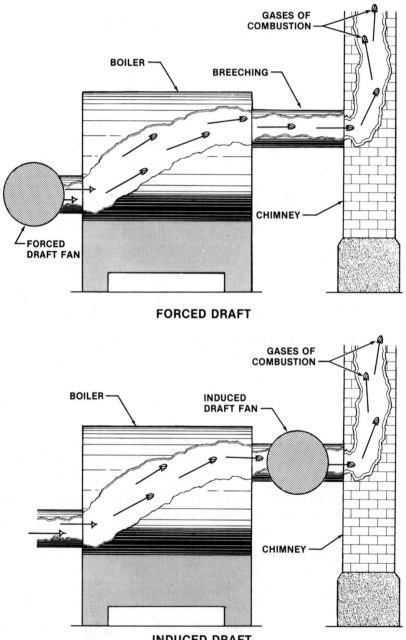

Figure 6-3. The location of the draft fan determines whether the draft is forced or induced.

inside and outside the chimney. In summer this same chimney cannot produce as much draft as in winter. With mechanical draft, the amount of draft produced can be controlled, resulting in more fuel burned and more steam produced.

When using forced draft, the proper chimney design is necessary to remove the hot gases of combustion from the furnace. With induced draft, chimney design is not as important because the gases of combustion are forced into the chimney. Consequently induced draft fans can to some degree help to overcome poor chimney design, which forced draft cannot. Some boilers have *combination forced and induced draft*, which uses both forced and induced draft fans. See Figure 6-4. Combination draft produces more draft and allows greater control by the boiler operator.

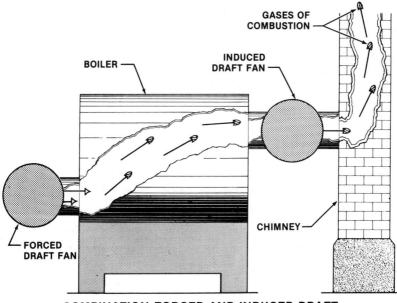

COMBINATION FORCED AND INDUCED DRAFT

Figure 6-4. Combination forced and induced draft uses power-driven fans located before and after the boiler.

MEASURING DRAFT

Special gauges are used to measure draft. *Draft gauges* are calibrated differently than other pressure gauges. A boiler has minimal draft pressure. Draft pressure is measured in inches or tenths of an inch of a vertical column of water.

A *manometer* is a simple draft gauge that consists of a bent glass tube shaped like a *U*. One end of the U-tube is open and the other end is connected to a flexible hose. Water is added to the U-tube until both legs have the water level at 0. The flexible hose is placed at the point where the draft is measured, such as the boiler breeching.

If the pressure in the breeching is less than in the atmosphere, the pressure will pull some of the air out of the hose, raising the water in the side of the manometer connected to the hose. This causes the water level to drop on the other side of the manometer. If the pressure in the breeching is more than outside pressure, the water in the side connected to the hose would be forced down, causing the water level in the open side to raise. See Figure 6-5. If the pressure in the breeching is less than atmospheric pressure, it is a *negative reading*. If the pressure is higher than atmospheric pressure, it is a *positive reading*.

CONTROLLING DRAFT

The method for controlling draft is determined by the boiler equipment used. The boiler operator must know how to control draft in a boiler. In a hand-fired boiler, the boiler operator has to adjust the openings in the ash pit door, the fire door, and dampers in the smoke pipe or breeching. This method requires practice to obtain the most efficient procedure. Consulting an experienced boiler operator in the plant will save training time.

Different controls are required for regulating draft in fuel oil or gas boilers, and in stokers. Boiler manufacturers

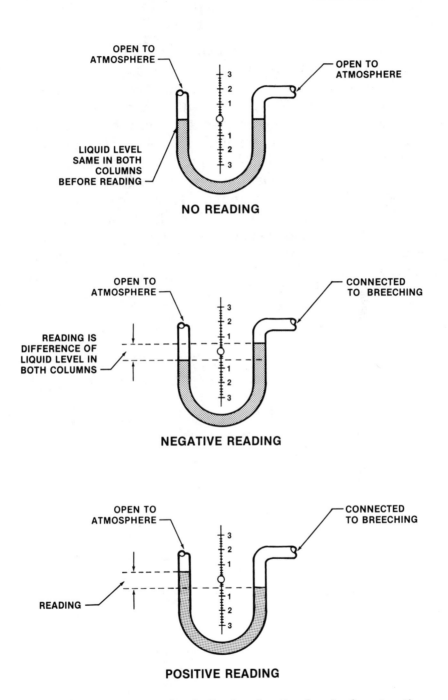

Figure 6-5. A manometer indicates by the level of water the amount of draft present at the point of measurement.

provide instructions for operating these controls and field representatives of these companies should be contacted with any questions concerning adjustment or operation.

Proper control of draft maintains high combustion efficiency. High combustion efficiency keeps smoke and air pollution to a minimum and produces the maximum amount of heat from the fuel. Too much draft may cause too hot a fire when burning coal. In addition, by removing gases of combustion from the furnace before the heat has been transferred to the water, heat escapes up the chimney. Too little draft will not provide enough air to efficiently burn the fuel. Improperly burned fuel results in carbon particles, soot, or smoke going up the chimney. In either case, the result of improper draft is wasted fuel.

Points to Remember

1. Draft is the difference in pressure between two points that causes air or gases to flow.
2. The amount of draft determines the rate of combustion or burning of the fuel.
3. The two types of draft are natural draft and mechanical draft.
4. Mechanical draft may be either forced, induced, or combination forced and induced.

Key Words

air for combustion
combination forced
  and induced draft
damper
draft
draft gauge
forced draft
gases of combustion

induced draft
manometer
mechanical draft
natural draft
negative reading
positive reading
power-driven fan
vacuum

BOILER WATER TREATMENT

7

Water used in the boiler must be treated for safety and efficiency. Minerals present in water can result in a build up of deposits that restricts water flow and causes overheating of boiler parts. Impurities float-

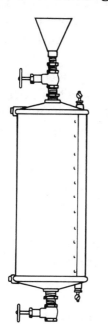

ing on the surface of the boiler water cause high surface tension resulting in priming and carryover.

Oxygen in the water can lead to corrosion and pitting of the boiler metal. The correct amount of chemicals must be added to the boiler water as required. If the concentrations of chemicals in the boiler water is too low, scale formation can result. If the concentration of the chemicals in the boiler water is too high, a bottom blowdown is required.

KEY TO ARROWHEAD SYMBOLS			
△ - Air	▲ - Water	▲ - Fuel Oil	⌂ - Air to Atmosphere
⬟ - Gas	△ - Steam	⬠ - Condensate	⬠ - Gases of Combustion

SCALE

Scale is formed by the collection of mineral deposits settling on the heating surfaces (tubes and boiler drums) of the boiler. All city water and most water from wells contain minerals. These minerals (salts) do not create problems as long as they are dissolved in water as a solution. However, water can only hold so much of these minerals. When concentrated, minerals settle out of the water and form scale (mineral deposits). Water that contains large quantities of minerals is *hard water.* Water that has small quantities of minerals is *soft water.*

When water is turned to steam, minerals present in the water are left in the boiler. The minerals then settle out of the water and form a scale on the boiler heating surfaces. An example of scale formation occurs in a teakettle. As the water is boiled away, scale collects on the bottom of the teakettle.

Scale causes problems in a boiler by acting as an insulator to slow down the transfer of heat to the water. This results in more fuel burned to generate the desired amount of steam. In addition, if the water cannot remove the heat, the heating surface will overheat.

For example, if a boiler tube without water in it was placed in the furnace of a boiler that was being fired, the tube would overheat and bend in a very short time. However, a boiler tube placed in the furnace or boiler with water passing around or through it (depending on the type of boiler) is not damaged. Damage to the tube will not occur if a sufficient amount of water removes the heat to prevent the boiler tube from overheating.

Overheating of heating surfaces reduces strength and results in blisters and bags formed in the metal and burned out tubes. Scale also narrows the inside of the tube and restricts the flow of water through it. This, along with the insulating effect of scale, contributes to overheating of the boiler tube. See Figure 7-1.

SCALE

Figure 7-1. Improper treatment of boiler water results in scale formation in boiler tubes. (*The Permutit Co. Inc.*)

Preventing Scale Formation

Scale formation is prevented by adding chemicals to the boiler water. These chemicals change minerals into *nonadhering sludge.* The sludge remains in the water and will not settle out. To remove the sludge, it is dropped to the bottom of the boiler and blown out through the bottom blowdown lines.

When a boiler is under a heavy load, it boils vigorously and the sludge in the water circulates rapidly. As the load drops off, the circulation slows down and the sludge settles to the lowest part of the boiler. The bottom blowdown line is used to remove sludge, and is opened regularly when the boiler is on light load. This permits blowing out sludge to prevent any possible scale formation.

OXYGEN IN BOILER WATER

Oxygen in the boiler water causes corrosion (rusting) and pitting of the boiler metal. Two methods are used to eliminate the problem of oxygen in boiler water. One method is to heat the water before it is put in the boiler to remove some of the air. The other method is to add chemicals that combine with oxygen to convert it into a harmless compound. These chemicals are called *oxygen scavengers.* Sodium sulfite is an oxygen scavenger that is commonly used to treat boiler water.

FEEDING CHEMICALS

In low pressure operation, chemicals for preventing scale and reducing oxygen are usually introduced into the boiler water by using a *bypass feeder.* The bypass feeder is installed on the discharge side of the boiler feedwater pump or vacuum pump. Hand-operated valves are used to control the amount of chemicals added by the bypass feeder. See Figure 7-2.

Valve 1 is open and valves 2 and 4 are closed. Valve 3 is opened slowly to make sure there is no pressure in the feeder. Pressure will be present only if valves 2 and 4 are leaking. When valve 3 is open, the chemicals are poured into the funnel.

Then valve 3 is closed, valves 2 and 4 are opened, and valve 1 is slowly closed. The water has to flow through the bypass feeder, feeding the chemicals into the boiler. This is sometimes referred to as *slug feeding* a boiler because the chemicals are moved into the boiler in slugs of water.

In some plants a more elaborate system for feeding chemicals is used. This system has a tank in which the chemicals are mixed with water. See Figure 7-3. A small pump on the bottom of the tank takes the chemical mixture from the tank and discharges it through pipelines directly to the boiler.

The pump is designed so its stroke may be adjusted to control the amount of chemicals being fed to the boiler. This method can be used to feed the boiler continuously, taking as long as 24 hours to empty the tank.

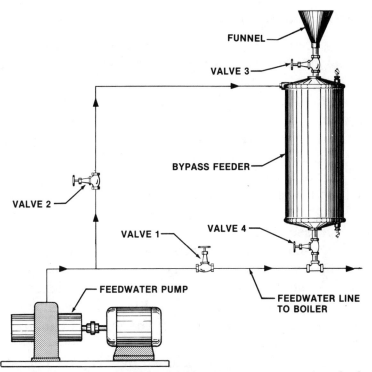

Figure 7-2. A bypass feeder adds oxygen scavenger chemicals to the boiler.

Figure 7-3. The chemical feed tank has an adjustable stroke pump that makes it possible to closely control the chemical feed to the boiler.

PRIMING AND CARRYOVER

Priming occurs when small particles of water are carried into the steam lines. Carryover occurs when large slugs of water get into the steam lines. Both of these problems can be caused by too high a water level in the boiler, by a high concentration of chemicals in the water, by impurities in the water that cause a high surface tension, or by opening a main steam valve too quickly. In addition, priming and carryover can cause water hammer, which could cause steam lines to burst.

Using a bottom blowdown reduces the water level and also decreases the concentration of chemicals. A *surface blowdown* may be needed if the water is foaming. The *surface blowdown valve* is located on the boiler directly at the NOWL. When the valve is opened, a mixture of steam and hot water is released. A surface blowdown removes any impurities that are floating on the surface causing the high surface tension.

High surface tension is shown by a rapid fluctuation of the level in the gauge glass. The glass fills and empties rapidly, indicating foaming in the boiler. Care in opening the steam valves prevents carryover caused by sudden pressure changes.

A bottom blowdown may be needed to reduce the boiler water level and to correct for high concentration of chemicals. To reduce the concentration of chemicals in the boiler water, the bottom blowdown should be followed by the addition of makeup water to dilute the chemicals to a safe level.

BOILER WATER TREATMENT

Boiler water treatment required for most low pressure boilers is minimal if the boiler recovers all or most of the condensate returns. The only makeup water needed is what is necessary to replace water loss from blowing down the gauge glass, water column, low water fuel cutoff, and from leaks. The less makeup water used, the less scale-forming salts that are introduced and the less chemicals that are needed. Fewer blowdowns will then be needed to remove the sludge.

The boiler operator must know which chemicals have been used to treat the feedwater and how often it has been treated. The chemicals used to treat boiler water depend on the minerals in the raw feedwater, which are determined by tests. If the chemicals required are unknown, contact a firm specializing in boiler water treatment, a field representative of the boiler manufacturer, or the local boiler inspector.

Many companies specialize in boiler water treatment. Contact two or three companies and have them send their representatives to take samples of the water used for makeup boiler water and condensate returns. The samples will be analyzed and a proposal detailing any boiler water problems will be submitted. The boiler operator should have a boiler water test kit to check boiler water and returns. The tests are simple and enable the operator to control the boiler water treatment.

BOILER WATER CONTAMINATION

The danger of contamination of boiler water by fuel oil occurs in plants that use No. 6 fuel oil. This fuel oil must be heated in order to pump it and heated again in order for it to burn. The steam heating coils may leak and cause fuel oil to get into the feedwater system with the condensate returns from the heaters. Fuel oil inside the boiler is very dangerous because it increases the surface tension, causing the water to foam.

The fuel oil also settles on the heating surfaces of the boiler and causes them to overheat. This can result in blisters and bags forming and the tubes burning out. If signs of fuel oil are found on the water side of the boiler, the boiler should be taken off-line immediately and cleaned.

To clean the boiler, it must be boiled out with caustic soda and then thoroughly washed. Consult a feedwater treatment company for specific chemicals to use in cleaning the boiler. Because of the danger of contamination of boiler water, it is a good practice to dump all fuel oil heater returns to waste. This prevents the boiler from being contaminated with fuel oil.

Points to Remember

1. All city water contains some minerals and scale-forming salts.
2. Water containing large amounts of minerals and scale-forming salts is called hard water.
3. Water containing small amounts of minerals and scale-forming salts is called soft water.
4. Scale build-up in boiler tubes reduces heat transfer and can result in overheating and weakening of the boiler tubes.
5. Chemicals added to the boiler turn minerals into a nonadhering sludge.
6. A bypass feeder is commonly used to add chemicals in a low pressure boiler.
7. Slug feeding is the feeding of chemicals in slugs of water into the boiler.
8. A chemical feed tank is used to add chemicals to the boiler continuously.
9. The nonadhering sludge is removed from the boiler using the bottom blowdown valves.
10. The bottom blowdown line should be opened regularly when the boiler is on a light load.
11. Oxygen is removed from the boiler water by heating

the water before it enters the boiler and by adding an oxygen scavenger to the boiler.

12. The oxygen must be removed from boiler water because it causes corrosion and pitting of boiler surfaces.
13. Priming is the carrying of small particles of water into the steam lines.
14. Carryover is when large slugs of water are carried over into the steam lines.
15. Priming and carryover can cause water hammer and rupture of steam lines.
16. A surface blowdown is used to remove impurities floating on the surface of the boiler water.
17. A bottom blowdown is used to reduce the chemical concentration in the boiler water.
18. Fuel oil in the steam and water side of a boiler can lead to overheating and burning out of the heating surface.
19. Fuel oil can be removed from the steam and water side of a boiler by boiling it out using caustic soda.
20. Boiler water treatment is an important part of safe and efficient operation of a plant.
21. A representative from a boiler water treatment chemical company can analyze and prescribe the treatment needed for a specific plant.
22. Contamination of boiler water by fuel oil can occur in plants using No. 6 fuel oil.

Key Words _____

bottom blowdown
bypass feeder
carryover
caustic soda
feedwater treatment
hard water
nonadhering sludge

oxygen scavenger
priming
scale
scale-forming salts
sodium sulfite
soft water
surface blowdown

BOILER OPERATION

8

Boiler room equipment varies depending on the type and size of the plant. The boiler operator is responsible for the safe operation of the boiler. Established operating procedures are developed through experience and manufacturers' specifications.

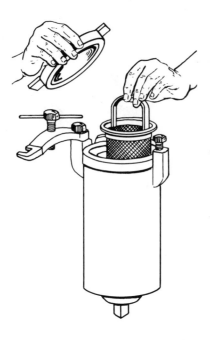

Boiler accessories must function properly to ensure safe and efficient boiler operation. The boiler operator must test and perform routine maintenance on boiler accessories as required. The ASME code has suggested procedures for the testing of accessories found on the boiler. Boiler operation data, testing, and maintenance procedures are recorded in the boiler room log.

KEY TO ARROWHEAD SYMBOLS			
△ - Air	▲ - Water	▲ - Fuel Oil	◠ - Air to Atmosphere
⚠ - Gas	△ - Steam	⚠ - Condensate	⚠ - Gases of Combustion

TAKING OVER A SHIFT

When taking over a shift, established procedures ensure that the boiler plant continues to operate safely and efficiently. Preliminary safety checks performed by the boiler operator identify any possible problems. The first thing a boiler operator should do when entering a boiler room to take over a shift is to check the water level on all boilers on the line generating steam. This is done by blowing

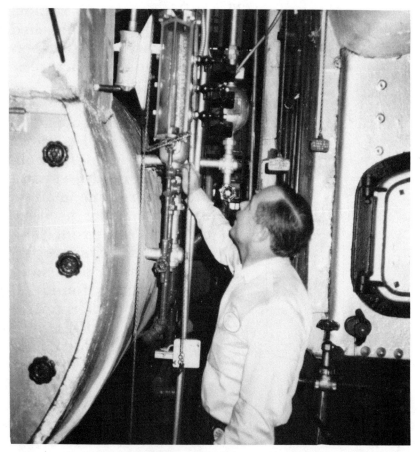

Figure 8-1. The water level in the gauge glass should drop out of sight immediately after the gauge glass blowdown valve is opened.

down the gauge glass. See Figure 8-1. Water should enter the glass quickly when the gauge glass blowdown valve is closed, which indicates the lines are free of sludge, sediment, or scale buildup.

After blowing down the gauge glass, the water column and low water fuel cutoff should be blown down. The boiler must be firing when the low water fuel cutoff is blown down to enable the boiler operator to check its operation. When the low water fuel cutoff is blown down, the burner should shut off. When the low water blowdown valve is closed, the boiler should light off again.

The boiler operator should then check the steam pressure and the condition of the fires. All running auxiliaries, including the fuel oil pumps, water pumps, fan, and burner, must be checked for proper temperature, pressure, and lubrication.

After these important preliminary safety checks have been performed, the operator should have routine duties to carry out. This routine helps to maintain consistency in performing various tasks. The following is a suggested routine during a shift.

1. Change over the fuel oil strainers and clean the one that has been in operation.

2. Shut down the burners one at a time and clean the burner tip or rotary cup, depending on the type of burner in the plant.

3. Remove the flame scanner with the burner firing. The burner should shut off as in a flame failure condition. Clean the flame scanner, replace it, and reset the programmer. Check for proper operation as the boiler goes through a firing cycle and lightoff. See Figure 8-2.

In addition to making preliminary safety checks and performing routine duties, the operator should determine when the safety valve was last checked and the accuracy of the gauges.

Figure 8-2. When the flame scanner is removed, the burner should shut off in a flame failure condition.

Boiler Start-up

When starting up a boiler, all equipment should be checked for proper operation. The following is a list of standard procedures to be completed during boiler start-up.

1. Check the water level in the boiler.
2. Check the main steam stop valve to make sure it is closed.
3. Check the fire side. Be sure there are no tools or rags left inside and no trace of fuel oil, gas, or fumes.
4. Open the boiler vent or the top try cock to vent

the air from the boiler during the warm-up.

5. Start the burner up and keep it in low fire to allow the boiler to warm slowly.

6. Watch the water level as the boiler is being warmed.

7. When the steam pressure gauge records pressure on the boiler, blow down the gauge glass, water column, and low water fuel cutoff.

8. Test the flame scanner to be certain of its operation.

9. When the boiler is a few pounds below the header pressure, slowly crack the main steam stop valve and allow the pressure to equalize. The main steam stop valve is opened slowly until it is wide open. See Figure 8-3. The boiler is now cut in on the line.

Figure 8-3. The main steam stop valve is opened slowly to prevent carryover and water hammer.

NOTE: The boiler should always be a little below the header pressure when cutting in on the line. This prevents carryover because steam flows into the boiler for a short time.

Boiler Shutdown

When taking a boiler off-line, make sure the remaining boilers can carry the load before proceeding with the following steps.

1. Secure (shut off) the fire on the boiler that is coming off-line.
2. Reduce the amount of forced draft to prevent the brickwork in the furnace from cooling too rapidly. Gradual cooling of the brickwork prevents spalling (flaking of the surface) from developing in the brickwork.
3. Watch the water level.
4. When the boiler has stopped steaming, close the main steam stop valve.
5. When the steam gauge shows about 1 or 2 pounds of pressure, open the boiler vent or top try cock to prevent a vacuum from forming.

NOTE: Do not close the main steam valve immediately because the safety valve will pop. The boiler furnace is hot and the heat in brickwork and tubes will continue to keep the boiler steaming for a short time.

Points to Remember

1. Check water levels, steam pressure, fires, and running auxiliaries immediately when coming on duty.
2. Clean strainers, burners, and other controls and equipment.
3. When starting a boiler, follow the start-up procedure by checking the water level, steam valves, boiler

vents, and controls.

4. Shut down a boiler by cutting off the fire, stopping the draft, and watching the steam until the valves can be closed to take the boiler off-line.

BLOWING DOWN A BOILER

The four reasons for giving a boiler a bottom blowdown are to

1. blow out sludge and sediment,
2. control high water,
3. control high chemical concentrations,
4. dump (empty) a boiler for cleaning, inspection, or repairs.

There should be pressure on the boiler at all times when it is blowing down except when being dumped. In addition, the boiler must be cooled before it is dumped because dumping a hot boiler results in damage to the boiler.

The following are precautions a boiler operator must take when blowing down a boiler.

1. Always check the water level before starting to blow down.

2. Try to blow down when the boiler is on a light load.

3. Never walk away from an open blowdown valve. The operator's hand must be kept on the valve until it is closed.

4. Open the blowdown valve slowly. Open it fully and then close it slowly. See Figure 8-4.

5. Never keep the blowdown valve open long enough for the water level in the boiler to drop out of sight in the gauge glass.

Some boilers are equipped with two valves on the blowdown line. A quick-opening valve and screw valve are

Figure 8-4. During bottom blowdown, the water level in the boiler must be closely monitored after opening the bottom blowdown valve.

most commonly used. Of the two valves the ASME code states that the quick-opening valve should be the one closest to the boiler. The quick-opening valve is a sealing valve. The screw valve is the blowing valve and is designed to withstand the wear and tear of blowing down. The quick-opening valve should be opened first and closed last.

LOW WATER CONDITIONS

Low water conditions can lead to overheating and a possible boiler explosion. After operating a boiler for a while, the operator knows where the normal operating water level (NOWL) is in the boiler. The NOWL varies in different boilers but is generally about one-half of a gauge glass. As the water level starts to fall below this one-half of a gauge glass, the boiler starts to develop a low water condition. If water can be seen in the gauge glass or if water comes out of the bottom try cock when it is opened, water can be added to the boiler safely.

NOTE: The lowest visible part of the gauge glass must be 2 " to 3 " above the highest heating surface. As long as the heating surface is covered with water, it is safe to add water to the boiler. If water cannot be seen in the gauge glass or if steam comes out of the bottom try cock, *do not attempt to add water to the boiler.* The burner must be secured, cooled slowly, and the boiler inspector notified. The boiler is then inspected for any damage caused from overheating. This is the same procedure that is used when the boiler drops (melts) a fusible plug.

WARNING: Never add water to a boiler if water cannot be seen in the gauge glass or if steam comes out of the bottom try cock when opened. Adding water could cause a boiler explosion.

FURNACE EXPLOSIONS

Furnace explosions are caused by a buildup of combustible gases or vapors in the furnace. This buildup could be the result of a leaking gas or fuel oil valve or a flame failure not handled properly.

To avoid a furnace explosion, if any odor of gas is detected the burner should be shut down and not be allowed to light off. All gas lines should be checked for leaks by applying a mixture of soapy water with a paint brush. Gas leaks will show up as bubbles. Any leaks must be repaired.

Another possible cause of a furnace explosion is leaking fuel oil in the furnace. Before lighting off a boiler, the furnace should be checked for signs of fuel oil. If fuel oil is found, it must be cleaned up and the source of the leak found and repaired.

If a flame failure occurs and the flame scanner has shut off the burner, the furnace must be examined for signs of fuel before lighting off. The furnace must then be purged to remove any fuel oil vapors present in the furnace.

One method of purging is to shut off the gas to the pilot, disconnect the spark from the ignition transformer, and shut off the fuel oil valve. Then start the programmer and let it go through a cycle. The burner motor will run and the fan will blow out any buildup of fuel oil vapors. Then open the fuel oil valve, reconnect the spark, open the gas valve, and allow the burner to go through a firing cycle.

Never turn the programmer ahead to bypass a purge cycle in order to start a burner that has misfired. This could cause additional unburned fuel oil to collect in the furnace. This fuel oil, as it vaporizes, could lead to a serious furnace explosion.

Points to Remember _____

1. The bottom blowdown valve is used to dump (empty) a boiler for repairs or inspection, remove sludge and mud from the boiler, control high water, and regulate chemical concentration of the boiler water by removing some of the high mineral water in the boiler.

2. The operator's hand must be kept on the blowdown valve when the valve is open. Operate the blowdown valve slowly and watch the water level while blowing down the boiler.

3. Secure the boiler (shut off the fire) if water cannot be seen in the gauge glass or if steam comes from the bottom try cock when it is opened.

4. The boiler must be inspected before operating it

again if it was shut down because of low water. This may prevent an explosion.

5. If any odor of gas or fuel oil is noticed in the furnace, shut off all fire-lighting devices and check for the source of the odor. The operator should watch for leaking lines, stuck valves, and defective lighting devices.

6. Purge the furnace before attempting to light a fire in gas or fuel oil fired boilers.

PREPARING FOR BOILER INSPECTION

Before a boiler can be inspected, it must be taken off-line. Once a boiler is off-line, the following safety checks must be made.

1. The main steam stop valve or valves (some boilers have two) must be closed and *tagged out.* To tag out a valve means to mark it so it will not be opened by mistake. Some plants attach a sign to the valve wheel that reads "Danger! Man in boiler—do not open."

2. The boiler vent or top try cock should be checked to see that it is open. This ensures there is no vacuum inside.

3. The feedwater line to the boiler must be closed and tagged out. If there is an automatic city water makeup valve, it must be secured also.

The boiler is now allowed to cool slowly. When the boiler is cool enough to dump, use the bottom blowdown valve to empty the boiler. After the boiler is empty, close and tag out the boiler bottom blowdown valve.

NOTE: A boiler is cool enough to dump when the heating surface is cool enough to touch. Never dump a boiler that is hot because all the sludge and sediment may bake on the heating surfaces, which is very difficult to remove.

As soon as the boiler has been dumped, open the hand-holes, remove the manhole cover, and thoroughly flush and wash out the water side. Do not dump a boiler unless it can be washed out immediately. If a boiler is dumped and not flushed right away, the sludge and sediment air dry on the heating surfaces, making it extremely difficult to clean.

As the water side is cleaned, look for signs of scale, pitting, or fuel oil. After cleaning the water side, the fire side must be inspected and cleaned thoroughly. See Figure 8-5. Remove all soot and carefully examine the entire fire

Figure 8-5. The fire side of the boiler is inspected for blisters or bags on the heating surface.

side. Look for signs of blisters or bags on the heating surface. Check the condition of the brickwork and make any
needed repairs.

Ask the local boiler inspector what preparations must
be made for the inspection. Very often the inspector wants
all of the plugs removed at the water column and the low
water fuel cutoff controls opened so the inside float chamber can be inspected. Fusible plugs must be replaced.
After both the fire side and water side of the boiler have
been cleaned, notify the inspector that the boiler is ready
for inspection.

Inspection dates are usually the same every year. Contact the inspector several weeks in advance to set a date
for inspection. This prevents having the boilers down for
an unnecessary period of time.

When the inspector is in the plant, have anything
ready that may be needed. For example, the inspector may
need a ladder and a low voltage droplight. Have someone
to assist if necessary, which will save time. If the operator
notices anything that might affect the safety of the boiler
operation, the inspector should be told. Follow any recommendations the inspector makes.

HYDROSTATIC TEST

A hydrostatic test is a water pressure test used on a boiler
to check for leaks or damage due to a low water condition
or done after any extensive repairs. The boiler inspector
may ask for a hydrostatic test if there is any doubt about
the boiler being able to carry its rated pressure safely after
the inspection has been completed.

In order to do a hydrostatic test on a boiler, the boiler
must be completely filled with water. To perform a hydrostatic test, the following operations must be carried out.

1. If the water column has a whistle valve, it
must be removed and plugged.

2. The main steam stop valve must be closed.

3. The safety valve or valves must be removed and blank flanges installed, or the safety valves must be gagged. (A gag is a clamp that prevents the valve from popping open without damaging the valve.)

4. The boiler vent must remain open until water comes out and then is closed.

5. Pressure on the boiler is now brought up to 1½ times the MAWP. (The pressure must be under control so that it does not exceed this pressure by more than 10 pounds.)

Watch the water temperature when filling the boiler. Water that is too cold may cause the boiler to sweat and leaks will be hard to find. Water that is very hot may flash into steam when it enters the boiler. After completing the test satisfactorily, the whistle valve is replaced and the safety valve gags removed or the valves replaced.

Points to Remember

1. Tag out a boiler being inspected on steam lines, water lines, and fire controls.
2. Have all tools and equipment ready for the inspector, such as a low voltage droplight, ladders, and tools.
3. Hydrostatic testing a boiler for leaks and strength by water pressure requires filling the boiler with water to the top and then applying pressure to the water. Be sure all valves are closed and fastened so the water cannot get into the steam lines or damage the relief valves.
4. Be sure the water is at room temperature or slightly above when filling the boiler for a test. This will avoid sweating due to cold water that could hide a real leak.
5. After the hydrostatic test, recheck all valves and return them to their normal operating condition.

LAYING UP A BOILER

Boilers that are to be out of service for any length of time should be laid up to prevent their deterioration. The two methods of laying up a boiler are *wet lay-up* and *dry lay-up*. Regardless of the method used, the boiler must be thoroughly cleaned on both fire and water sides. All traces of soot must be removed. In a coal-fired boiler, all traces of coal and ash must be removed. Coal, ash, or soot contains sulfur. When combined with water, sulfur reacts to form sulfuric acid, which corrodes boiler metal.

A wet lay-up is done after the boiler has been thoroughly cleaned on both fire and water sides. Then the boiler should be closed up. New gaskets are used on all handholes and on the manhole. Be sure that nothing is left inside the boiler after cleaning. The boiler is then filled to the top with warm, chemically treated water to cut down on corrosion and oxygen pitting.

A dry lay-up is recommended by the ASME code for boilers that will be out of service for a long time or a boiler that is in danger of freezing. After both fire and water sides of the boiler have been thoroughly cleaned, the water side must be carefully dried. Any moisture left on the metal surfaces causes corrosion.

The handholes are replaced using new gaskets. Trays of quick lime (2 pounds per 1,000 gallon capacity) or silica gel (10 pounds per 1,000 gallon capacity) should be placed on the water side of the boiler. The manhole cover is then replaced using a new gasket. All valves and connections are closed to prevent any moisture from entering. The trays of chemicals should be examined at regular intervals and replaced when necessary.

BROKEN GAUGE GLASS

A broken gauge glass on a boiler under pressure is replaced using the following procedure.

1. Secure (close) the water valve to the gauge glass.
2. Secure the steam valve to the gauge glass.
3. Open the gauge glass blowdown valve.
4. Remove the gauge glass nuts.
5. Remove the broken glass and the washers.
6. Get a new gauge glass and new washers.
7. Put the gauge glass and the washers in place.
8. Hand tighten the gauge glass nuts and then give them a one-quarter turn with a wrench.
9. Crack the steam valve to the gauge glass to allow the glass to warm up slowly.
10. Then open the steam valve completely.
11. Now open the water valve to the gauge glass completely.
12. Close the gauge glass blowdown valve and check for leaks.

If a gauge glass must be cut to fit, allow ¼ " under the inside measurement to allow the glass to expand when it warms up. It is also important to use new gauge glass washers when replacing the glass.

CLEANING THE GAUGE GLASS

A gauge glass that is dirty on the inside must be cleaned to obtain accurate readings. The following steps should be followed when cleaning a gauge glass.

1. Secure water and steam valves to the gauge glass.
2. Open the gauge glass blowdown valve and make sure the valves are not leaking steam or water.
3. Remove the gauge glass nuts and gauge glass.
4. Use a cloth wrapped around a wooden dowel to clean the inside of the gauge glass. Never use an object that would scratch the glass. A scratch on the inside of the glass will start the steam cutting the glass, causing it to break.

5. Use new gauge glass washers when replacing the glass.

6. Follow normal recommended boiler warm-up procedures.

LEAKING GAUGE GLASS WASHERS

Leaking gauge glass washers result in a false water level reading. In addition, leaking and worn gauge glass washers can cause the gauge glass to wear quickly and eventually break. Apply pressure on the gauge glass washers by tightening the gauge glass nuts carefully. If tightening the nuts does not stop the leaking, the washers should be replaced. Safety goggles should be worn when working around the gauge glass. A supply of gauge glass washers should always be kept on hand.

CORRECTING A STEAMBOUND PUMP

A pump becomes steambound when the water it is pumping becomes too hot and turns to steam. Because a pump is designed to handle liquids, it does not work with steam or air. A steambound pump is usually caused by a faulty steam trap that allows steam to blow into the return lines. If this condition is not corrected promptly, damage to the pump or the motor driving it can occur.

To correct the steambound condition, the water fed to the pump must be cooled. If the pump is taking its suction from a condensate return tank or feedwater heater, city water can be added through the makeup system to drop the temperature. The traps returning water to the vacuum tank or condensate return tank should be checked and repaired. See Chapter 4.

A vacuum pump that is receiving condensate that is too hot will result in a loud banging noise. Cool the water by adding city water through the makeup system. If this does not cause the pump to start pumping again, cold water should be poured over the pump. When cooling the pump, be sure to pour water on the pump and never on the motor.

TESTING THE LOW WATER FUEL CUTOFF

The low water fuel cutoff is tested by blowing down or by an evaporation test. Blowing down the low water fuel cutoff is done with the burner firing. The blowdown valve is opened on the low water fuel cutoff. Water and steam rushing out will clean out any sludge and also allow the float to fall. This shuts off the fuel valve just as if there were a low water condition in the boiler. The low water fuel cutoff should be blown down daily or every shift.

The evaporation test is used to test the low water fuel cutoff by actually causing a low water condition to occur. This test is done by securing the vacuum pump or the feedwater pump so no water is supplied to the boiler and the water level gradually drops. *NOTE:* If the boiler has an automatic city water makeup feeder it must also be secured. The evaporation test is more accurate than blowing down the low water fuel cutoff. Under normal conditions the water level in a boiler will drop slowly and the float in the low water fuel cutoff may stick.

However, when a blowdown is used to test the low water fuel cutoff, the force from the water and steam will jar the float and make it drop in the test even though the valve may stick under normal conditions. The burner should shut off when water is still visible in the gauge glass. If the low water fuel cutoff fails to cut off the burner, someone should attend the boiler as long as it is firing until it can be taken off-line and checked.

MAINTAINING A BOILER ROOM LOG

A boiler room log is used to record information regarding operation of the boiler during a given period of time. See Figure 8-6. Some plants maintain a log for every 8-hour shift. Other plants maintain a log for a 24-hour period.

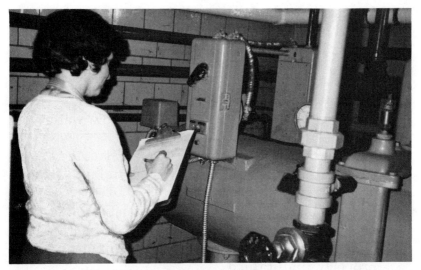

Figure 8-6. The boiler room log is used to accurately record information regarding the operation of the boiler.

Boiler data that is commonly recorded on a boiler room log includes

1. steam pressure,
2. boilers on line,
3. condensate return temperature,
4. feedwater temperature and pressure,
5. fuel oil temperature and pressure,
6. outside temperature.

Boiler operator duties performed commonly recorded on a boiler room log include

1. gauge glass blowdown,
2. water column blowdown,
3. low water fuel cutoff blowdown,
4. flame scanner testing,
5. safety valve testing,
6. boiler blowdown,
7. fuel on hand (gas meter readings for gas-fired boilers),
8. fuel consumed during shift,
9. maintenance work.

Maintaining a boiler room log allows the operator to evaluate the past performance of the boiler. In addition, boiler room log information can be useful in determining the cause of a malfunction and/or predicting a possible problem.

Points to Remember

1. Fill the boiler to the top with warm, treated deaerated water and close all valves when laying up for a short time.
2. For a long lay-up, trays of moisture-absorbent material are placed in the boiler.
3. Gauge glasses are extremely important and must be properly maintained.
4. Vacuum and feedwater pumps may become steambound if steam is present in the vacuum tank.
5. An evaporation test should be performed monthly to test the low water fuel cutoff.
6. A boiler room log is used to record information regarding the operation of a boiler.

Key Words

boiler lay-up
boiler shutdown
boiler start-up
bottom blowdown

furnace explosion
hydrostatic test
tagged out

HOT WATER HEATING SYSTEMS

9

Heating systems provide heat to designated areas by transporting heat energy generated in the boiler. The two types of heating systems are the steam heating system and the hot water heating system.

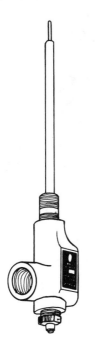

The difference in the two heating systems is the medium used to transport heat energy from the boiler to the area to be heated. Steam is used to transport heat energy in the steam heating system, and water is used to transport heat energy in the hot water heating system.

Boiler fittings (trim) are required to safely and efficiently transport steam or hot water used in a heating system.

KEY TO ARROWHEAD SYMBOLS			
△ - Air	▲ - Water	▲ - Fuel Oil	⋒ - Air to Atmosphere
⟁ - Gas	△ - Steam	⟁ - Condensate	⟁ - Gases of Combustion

HOT WATER HEATING SYSTEMS

Hot water heating systems transport heat by circulating heated water to a specific area. Heat is released from the water as it flows through the heating unit. After heat is released, the water returns to the boiler to be reheated and recirculated. See Figure 9-1.

Hot water heating systems produce heat more consistently than steam heating systems. The water in a hot water heating system remains in the lines at all times. The water in the heating unit lines heats and cools slowly, resulting in an even rate of heat production.

In the steam heating system, when pressure is lost steam leaves the the heating units, resulting in a more rapid loss of heat than in a hot water heating system. In addition, the steam heating system has a longer recovery

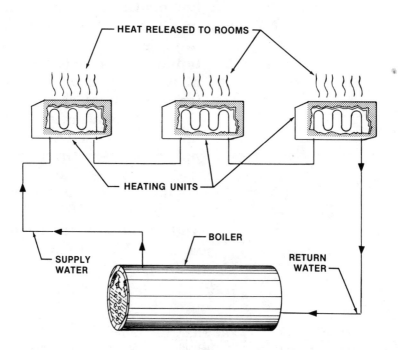

Figúre 9-1. Heating units release the heat of the supply water from the boiler as it circulates through the hot water heating system.

time in producing heat after the boiler is shut down than the hot water heating system.

More water (by weight) is required in a hot water heating system. Water conveys approximately one-fifth the amount of heat that steam conveys. To produce the same amount of heat, 50 times more water (by weight) is required. In addition, a typical hot water heating system usually operates at 170 °F; a steam heating system operates at 212 °F.

Hot water heating systems are designed to produce and regulate the circulation of hot water in a system. In the system, water is heated by the boiler in a central location. This equalizes the amount of heat loss as the water is distributed to the areas to be heated. The distribution of the heated water is determined by the type of hot water heating system used. The two types of hot water heating systems are the *natural circulation hot water system* and the *forced circulation hot water system*.

Natural Circulation Hot Water Heating System

A natural circulation (gravity feed) hot water heating system circulates hot water by using the difference in water density (weight per unit volume) of the hot water flowing to the heating units and the cool water flowing to the boiler. When water is heated, it increases in volume (expands) and loses weight. As water cools, it decreases in volume and gains weight. Water flow in the natural circulation hot water heating system is maintained by pressures resulting from the heating and cooling of the water in the system. See Figure 9-2.

Water is heated in the boiler (1). The water expands and becomes lighter, flowing up through the supply line (2) to the heating units (3). As the water is cooled in the heating units, the water becomes heavier and flows down through the return line (4) to the boiler.

The expansion tank (5) is located at the highest point

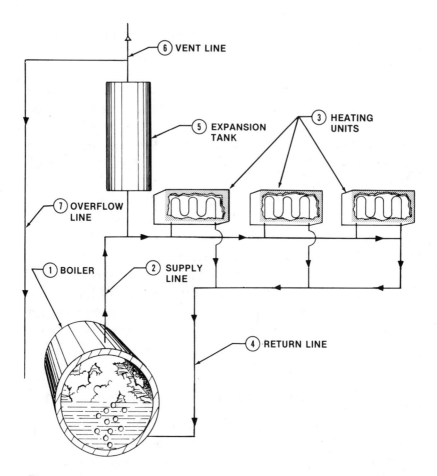

Figure 9-2. Heated water expands and loses weight to maintain circulation in the natural circulation hot water heating system.

in the system. The expansion tank functions as a relief valve, collecting excess water as the water is heated and its volume increases. The vent line (6) opens to the atmosphere and allows pressure to vary in the system as required. Excess water in the system is piped down an overflow line (7).

Natural circulation hot water heating systems became less popular as larger buildings required increased hot water flow rates. The larger diameter piping necessary for maintaining water flow increased installation costs, and as buildings became taller, natural circulation could not overcome the hydrostatic pressures present to maintain the necessary flow of hot water. Hydrostatic pressure is pressure exerted by water based on .433 psi per vertical foot. For example, water pressure at the bottom of a 10′ column of water is increased by 4.33 psi (10 × .433).

Forced Circulation Hot Water Heating System

The forced circulation hot water heating system circulates hot water by using *circulating pumps.* Circulating pumps are centrifugal pumps and vary in size and capacity, depending on the size of the plant. See Figure 9-3. Circulating pumps are designed for continuous or intermittent operation. Large plants use floor-mounted circulating pumps that operate continuously. Smaller plants most commonly use line-mounted circulating pumps that oper-

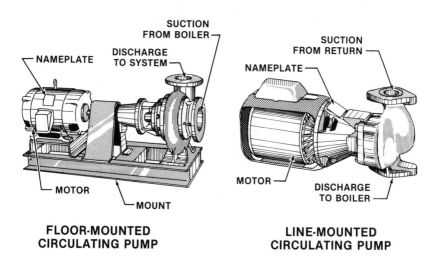

FLOOR-MOUNTED CIRCULATING PUMP

LINE-MOUNTED CIRCULATING PUMP

Figure 9-3. Circulating pumps operate continuously or intermittently in a forced circulation hot water heating system.

ate intermittently, depending on plant requirements. Specifications are listed on the nameplate located on the circulating pump.

The forced circulation hot water heating system is a closed system that has no vent to the atmosphere. See Figure 9-4. Water is heated inside the boiler (1). The aquastat (limit control) (2) senses the temperature of the water and controls the starting and stopping of the burner. The temperature-pressure gauge (3) indicates the temperature and pressure of the hot water in the system.

Water passes through the air control (4) and air present in the system is removed. The flow control valve (5) prevents natural circulation when the hot water is not being pumped through the system. The supply line (6) provides hot water to the heating units (7). Air vents (8) on the heating units allow the manual bleeding of trapped air in the heating unit. Trapped air in the heating unit prevents hot water from circulating through the heating unit, resulting in a cold heating unit.

Diverter fittings (9) allow the hot water to flow to the heating unit and continue to flow to the next diverter fitting. Water returns through the return line (10) to the circulating pump (11). Water is pumped back to the boiler for reheating. The compression tank (12) provides relief from pressure in the system caused by water expanding from heat. In addition, the compression tank collects air removed from the water by the air control and absorbs shock in the system caused by the starting or stopping of the circulating pump.

A second air control (13), located on the compression tank with a drain, is used for draining a compression tank that has filled with water. Without the air control, air is not provided to the compression tank during drainage, preventing water from flowing through the drain.

The compression tank valve (14) allows the compression tank to be drained without shutting down the system. The safety relief valve (15) relieves pressure in the system

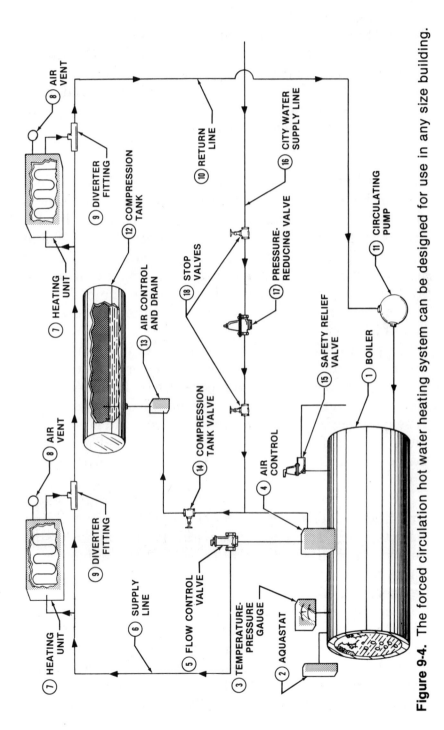

Figure 9-4. The forced circulation hot water heating system can be designed for use in any size building.

beyond the capacity of the compression tank. Water is added to the system by the city water supply line (16) and is reduced to 12 to 18 psi by the pressure-reducing valve (17). Stop valves (18) before and after the pressure-reducing valve allow servicing of the pressure-reducing valve without shutting down the system.

Water Temperature Control.   Water temperature control in the forced circulation hot water heating system is necessary to minimize boiler operation costs while providing sufficient heat. The boiler operator is responsible for controlling the temperature of the water in the boiler and the water circulating to the heating units in the system. The temperature of the water in the boiler can be controlled by aquastats or a remote temperature-monitoring system.

Aquastats are mounted on the boiler and control the firing of the burner to raise or lower boiler water temperature based on the temperature of the water in the boiler. The remote temperature-monitoring system uses a sensing device at a location away from the boiler. The temperature of the water in the boiler is raised or lowered based on the temperature at the remote location at a given time.

For example, in a hot water system used to heat a building, the sensing device mounted outside the building detects changes in outside temperature. When the outside temperature drops, a signal is relayed to the burner to raise the temperature of the boiler water. Raising the temperature of the boiler water allows the heating system to prepare for increased hot water demands caused by the drop in temperature outside the building. See Figure 9-5.

The sensing device (1) outside the building senses the temperature and sends a signal indicating the temperature to the master controller (2). The signal is relayed to the submaster controller (3). The submaster controller

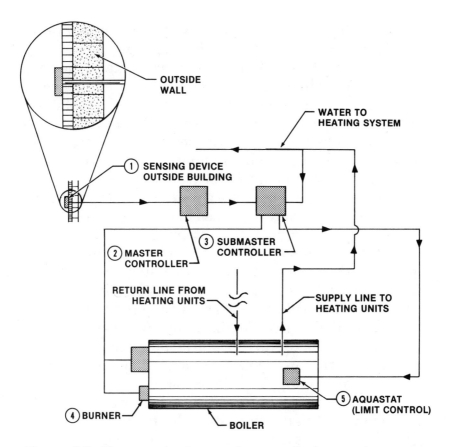

Figure 9-5. The remote temperature-monitoring system uses changes in temperatures outside the building to control the firing rate of the burner.

receives the temperature reading and compares it with the temperature of the water circulating in the heating system.

If the temperature of the water circulating in the heating system is below the desired range, the submaster controller sends a signal to the burner (4) to control the ON/OFF firing and the firing range of the burner as necessary. An aquastat (5) on the boiler can be used as a high-limit temperature control, which will secure the burner if the temperature of the water circulating in the heating system exceeds the desired range.

The temperature of the water circulating in the heating system can be controlled by a *three-way mixing valve*, starting or stopping the circulation pumps, or a combination of both. A three-way mixing valve automatically blends water returning from the heating units with supply water from the boiler. If the temperature of the return water is too low, supply water from the boiler is added. If the temperature of the return water is still in the desired range, no supply water is added and the water continues to circulate in the heating unit circuit.

Starting and stopping the circulating pumps determines the flow rate of the water circulation in the heating unit circuit. If the pump is stopped, hot supply water will not release its heat as quickly, causing the temperature of the system to remain high. Using a three-way mixing valve in addition to starting and stopping the circulating pump offers the best control of water temperature in the heating unit circuit. See Figure 9-6.

The three-way mixing valve (1) has two inlet valves and one outlet valve to control the temperature of the water going to the heating unit (2). The circulating pump (3) controls the amount of water flowing through the heating unit. The circulating pump is turned on and off as necessary by the thermostat (4). The water continues to recirculate in the heating unit circuit unless heated supply water from the boiler is introduced. The amount of supply water introduced is equal to the amount of water returning to the boiler from the heating unit circuit.

HOT WATER BOILERS

Hot water boilers are similar in design and construction to steam boilers. Hot water boilers can be classified as *low pressure* or *high pressure.* See Figure 9-7. Hot water boilers operating with a 250 °F water temperature and 160 psi

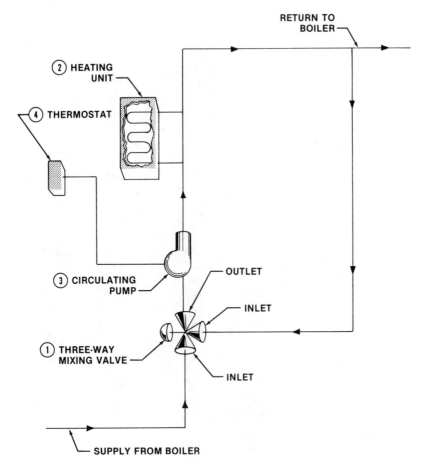

Figure 9-6. In the heating unit circuit, the circulating pump is controlled by the thermostat, and the amount of supply water is controlled by the three-way mixing valve.

water pressure or less are classified as low pressure. Hot water boilers operating with a water temperature exceeding 250 °F and/or a water pressure exceeding 160 psi are classified as high pressure.

Steam boilers are classified as high pressure when the steam pressure exceeds 15 psi. All boilers, whether hot water or steam, are manufactured in conformance with

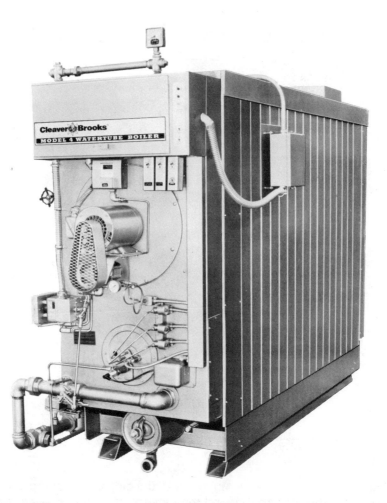

Figure 9-7. Low pressure hot water boilers operate with a 250°F water temperature and 160 psi water pressure or less. (*Cleaver-Brooks*)

Section I or IV of the ASME Boiler and Pressure Vessel Code. The specifications for operation of a boiler are found stamped or on a nameplate permanently mounted on the boiler. See Figure 9-8.

Low pressure boilers used to be manufactured for use as hot water or steam boilers. The fittings installed determined whether the boiler was for use with hot water or

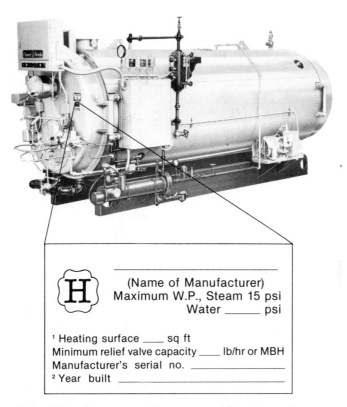

Figure 9-8. All boilers must have specifications for operation stamped or on a nameplate mounted on the boiler. (*Cleaver-Brooks*)

steam. Boilers were commonly stamped MAWP steam—15 psi and MAWP water—30 psi. Many low pressure hot water boilers used today are designed to operate at higher temperatures and pressures (up to 250 °F and 160 psi) to meet the demands of large modern buildings.

The three types of hot water boilers are fire tube, water tube, and cast iron sectional. Most low pressure hot water boilers are either fire tube or cast iron sectional boilers. Water tube hot water boilers are generally high pressure boilers. In a high pressure hot water system, high temperature, high pressure water is circulated from a central boiler to various locations. See Figure 9-9.

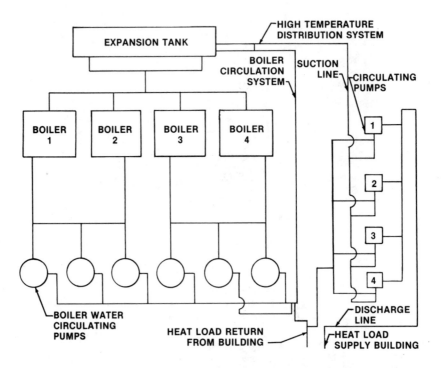

Figure 9-9. High temperature, high pressure hot water systems can use centrally located boilers to heat remote locations.

Boiler Fittings

Boilers fittings (trim) are required on the boiler to ensure the safe and efficient operation of the boiler. Hot water boilers require different fittings than steam boilers. For example, in a hot water system the aquastat starts and stops the burner on water temperature demand. In a steam system, the pressure control starts and stops the burner on steam pressure demand. The fittings required on a forced circulation hot water boiler are the safety relief valve, temperature-pressure gauge, air control, flow control valve, pressure-reducing valve, compression tank, aquastat, and diverter fittings.

Safety Relief Valve. The safety relief valve on the hot water boiler protects the boiler from exceeding the MAWP. The hot water boiler safety relief valve is rated in Btu's (water temperature) relieved per hour. In a steam boiler, the safety valve is rated in pounds of steam relieved per hour. The safety relief valve consists of the inlet, outlet, body, sealing diaphragm, spring, and hand-testing lever. See Figure 9-10.

The safety relief valve must be built according to the ASME code and have specifications identified on a data plate. The data plate includes the ASME symbol, model

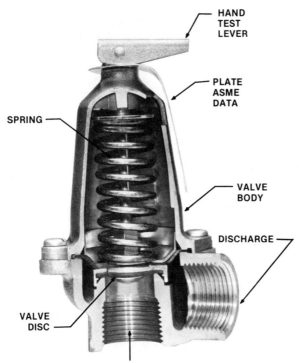

Figure 9-10. The safety relief valve prevents the boiler from exceeding its MAWP. (*Bell & Gossett Co.*)

number, part number, popping pressure, inlet diameter, capacity in Btu's per hour, and the manufacturer's name.

The safety relief valve should be located on the highest part of the boiler and should discharge to an open drain in the boiler room. The popping pressure of a safety relief valve on a low pressure hot water boiler cannot exceed 160 psi. Section VI of the ASME code recommends safety relief valves be manually tested every 30 days. In addition, safety relief valves should be pressure tested annually, preferably at the beginning of the heating season.

Temperature-Pressure Gauge. The temperature-pressure gauge indicates the temperature and pressure of the water leaving the boiler. See Figure 9-11. The temperature-pressure gauge is installed on the front or top of the boiler for easy viewing by the operator.

A temperature-pressure gauge has two scales. The top scale is the altitude pressure gauge and has two needles.

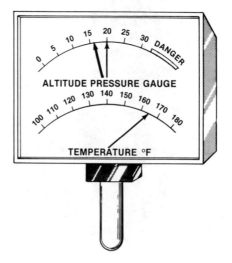

Figure 9-11. The temperature-pressure gauge indicates the pressure above the hydrostatic pressure required to pump hot water in the system.

One needle is adjusted by the operator to indicate what pressure is required to overcome hydrostatic pressure (.433 per vertical foot) to pump hot water through the system.

The other needle is the pressure in psi of water leaving the boiler. The bottom scale indicates the temperature of the water leaving the boiler. Hot water boilers usually have 12 to 18 psi on the boiler at all times. The pressure varies depending on the height and distance water must be pumped.

Air Control. Air controls remove air from the hot water boiler system. See Figure 9-12. Air prevents water from circulating in the heating units and causing corrosion in the system. Air controls are installed on the outlet line leaving the boiler and on the compression tank. Air re-

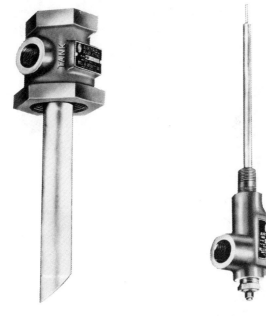

BOILER AIR CONTROL   COMPRESSION TANK
                            AIR CONTROL

Figure 9-12. Air controls are installed on top of the boiler to remove air from the supply line and on compression tanks to facilitate draining of the tank. (*Bell & Gossett Co.*)

moved from the water leaving the boiler is diverted to the compression tank. Water continues to flow to the flow control valve.

Flow Control Valve. The flow control valve opens and closes automatically to prevent natural circulation in the hot water heating system when the circulating pump is not running. The flow control valve is designed for either horizontal or vertical installation and is located on the outlet line of the hot water boiler. See Figure 9-13.

The flow control valve functions like a check valve. When the circulating pump is running, the flow control valve opens to allow water to flow out of the boiler to circulate in the system. When the circulating pump is not running, the flow control valve closes to prevent water from circulating by natural circulation. If a circulating pump fails, the flow control valve can be manually opened to provide hot water to the heating units through natural circulation until the circulating pump is repaired.

HORIZONTAL INSTALLATION   **VERTICAL INSTALLATION**

Figure 9-13. Flow control valves prevent natural circulation when the circulating pump is not running. (*Bell & Gossett Co.*)

The flow control valve can be disassembled and cleaned without breaking the pipe connections. See Figure 9-14. *NOTE:* Pressure must be off the system before the flow control valve is disassembled.

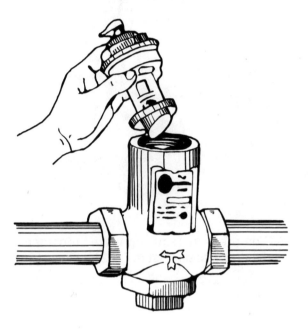

Figure 9-14. Pressure must be off the system when the flow control valve is disassembled for cleaning. (*Bell & Gossett Co.*)

Pressure-reducing Valve. The pressure-reducing valve reduces incoming city water pressure to approximately 12 to 18 psi for use in the hot water boiler system. See Figure 9-15. The pressure-reducing valve is equipped with stop valves and a bypass line. This permits the valve to be serviced without shutting down the system. The strainer in the pressure-reducing valve must be inspected periodically to ensure proper city water flow. Inspection should be more frequent if water containing scale-forming salts is used in the hot water boiler system.

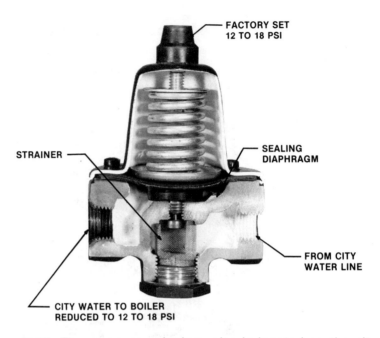

FACTORY SET
12 TO 18 PSI

STRAINER

SEALING
DIAPHRAGM

FROM CITY
WATER LINE

CITY WATER TO BOILER
REDUCED TO 12 TO 18 PSI

Figure 9-15. The pressure-reducing valve is located on the city water line and reduces the pressure to 12 to 18 psi. (*Bell & Gossett Co.*)

Compression Tank. The compression tank absorbs water pressure in the hot water boiler system. See Figure 9-16. Because the forced circulation hot water heating system is a closed system, no venting is provided. Water expands when heated, increasing pressure. The compression tank serves as a relief device, absorbing the variations in water pressure in the system caused by water temperature changes. The compression tank also absorbs shock to the system caused by the starting and stopping of the circulating pump.

Compression tanks are normally half full of water. This allows the level in the tank to increase or decrease while still maintaining water in the system. If the compression tank is allowed to fill with water, the pressure

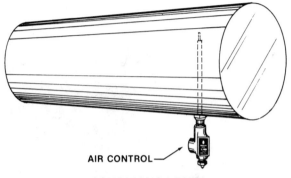

AIR CONTROL

COMPRESSION TANK

Figure 9-16. The compression tank collects water and pressure caused by the expansion of water when heated.

in the boiler will increase. This causes the safety relief valve to open and protects the system from exceeding the MAWP.

If the safety relief valve opens, the compression tank must be secured from the system. The compression tank is then drained and refilled to the proper level. Larger compression tanks are equipped with a gauge glass for determining the compression tank water level.

Aquastat. The aquastat (limit control) controls the starting and stopping of the burner by sensing the temperature of the water in the hot water boiler. See Figure 9-17. The aquastat consists of a mercury-tube switch that is actuated by a temperature-sensitive coil that moves with water temperature variation. The mercury-tube switch completes or breaks an electrical circuit to turn the burner on or off.

Aquastats usually have a temperature differential preset by the manufacturer. For example, a temperature differential of 160 °F and 180 °F starts the burner when the boiler water temperature drops to 160 °F and con-

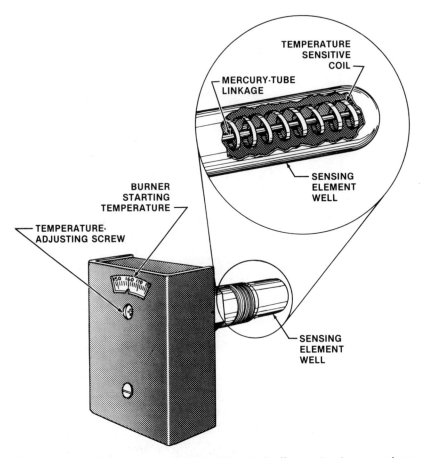

Figure 9-17. The aquastat is sensitive to boiler water temperature and controls the stopping and starting of the burner as required.

tinues to fire until the boiler water temperature reaches 180 °F. A built-in temperature differential eliminates the possibility of the burner continuously starting and stopping (hunting) at small temperature variations.

Some hot water boilers are equipped with a separate aquastat for *low limit,* which is the water temperature at which the burner is fired and for *high limit,* which is the water temperature at which the burner is shut off.

Diverter Fittings. Diverter fittings are used to provide water to a heating unit while maintaining the required flow of hot water to the next heating unit. Diverter fittings make it possible to use a single pipe to act as a supply to and return from the heating unit to the supply line. See Figure 9-18.

The number of diverter fittings required is determined by the location of the heating units in relation to the supply line. In addition, in heating units above the supply line where resistance to circulation is high, two diverter fittings may be required.

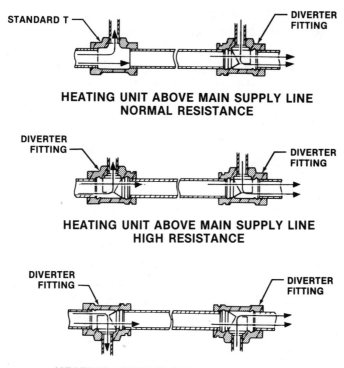

STANDARD T

DIVERTER FITTING

HEATING UNIT ABOVE MAIN SUPPLY LINE
NORMAL RESISTANCE

DIVERTER FITTING

DIVERTER FITTING

HEATING UNIT ABOVE MAIN SUPPLY LINE
HIGH RESISTANCE

DIVERTER FITTING

DIVERTER FITTING

HEATING UNIT BELOW MAIN SUPPLY LINE
NORMAL RESISTANCE

Figure 9-18. Diverter fittings are used to direct water flow to heating units and return lines off the same supply line. (*Bell & Gossett Co.*)

Points to Remember

1. Hot water heating systems require special hot water boiler accessories.
2. The two types of hot water heating systems are the natural circulation hot water system and the forced circulation hot water system.
3. In the natural circulation hot water heating system, water is circulated by the difference in water density between hot and cold water.
4. Hydrostatic pressure is pressure exerted by water based on .433 psi per vertical foot.
5. Forced circulation hot water heating systems use circulating pumps to circulate hot water.
6. Hot water boilers can be classified as low or high pressure.
7. The safety relief valve protects the hot water boiler from exceeding the MAWP.

Key Words

air control
aquastat
boiler fitting
circulating pump
compression tank
diverter fittings
forced circulation
   hot water system
heating unit
hot water boiler
hot water heating system

hydrostatic pressure
natural circulation
   hot water system
pressure-reducing valve
remote temperature-
   monitoring system
safety relief valve
steam heating system
temperature-pressure
   gauge
three-way mixing valve

BOILER ROOM SAFETY

10

Boiler room safety procedures must be exercised at all times by the boiler operator. Accidents can occur as a result of not following safety procedures or because of boiler equipment failure. If an accident

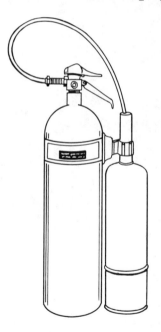

occurs, the boiler operator must act quickly. Established emergency procedures will reduce the possibility of additional injuries and/or damage to equipment. All boiler room accidents must be reported regardless of their nature.

Fuels used in the boiler room are combustible and present a fire hazard. The boiler operator is responsible for fire prevention in the boiler room. Boiler room safety is achieved by following safety rules.

KEY TO ARROWHEAD SYMBOLS			
△ - Air	▲ - Water	▲ - Fuel Oil	⌂ - Air to Atmosphere
⚠ - Gas	△ - Steam	⚠ - Condensate	⚠ - Gases of Combustion

BOILER ROOM ACCIDENTS

Boiler room accidents can occur at any time. Even though safety precautions are followed, the possibility of an accident still exists. See Figure 10-1. Injuries resulting from accidents must be handled quickly and intelligently. A boiler operator who is familiar with the equipment and plant will know how to react in an emergency.

Figure 10-1. A boiler explosion can result if safety precautions are not followed.

All injuries no matter how minor should be treated promptly. Serious injuries require notifying qualified personnel. All accidents should be reported regardless of their nature. Serious problems can occur regarding insurance claims if complications arise as the result of an accident that was not reported or put on file.

Accident reports include the following information: date, time and place of accident, immediate superior, name of injured person, nature of injury, what injured person was doing at time of accident, and cause of accident. Accident reports are also used to document plant safety records.

BOILER ROOM FIRE PREVENTION

Boiler room fire prevention procedures are necessary because of the combustible nature of the materials used in the boiler room. The boiler operator must know the procedure to be used in sending for the fire department or sounding the fire alarm.

The person who sends for help should make sure another plant worker is available to direct the fire fighters to the right location when they arrive. In addition, the boiler operator must know the location of fire alarm boxes and stations in or near the boiler room.

Combustible materials burn readily and require special handling by the boiler operator. The boiler operator must know what is necessary to start and sustain a fire in order to know how to put the fire out.

Fuel (combustible material), heat, and oxygen are required to start and sustain a fire. The fire will go out when any one of these is removed. See Figure 10-2. Fuel may be fuel oil, wood, paper, textiles, or any other material that burns readily. If the fuel supply is cut off or the fuel is burned up the fire will go out.

The fuel must be heated to its ignition temperature. If the burning material is cooled below its ignition temperature the fire will go out. Oxygen is required to support the combustion process. If the oxygen supply is cut off by smothering the fire will go out.

Since the main ingredient is the combustible material, waste or oil rags must be stored in safety containers and volatile liquids in safety cans. By maintaining careful control of the combustible materials in a boiler room, the danger of a fire hazard is reduced. Local fire departments have trained inspectors that inspect buildings and factories for possible fire hazards and required firefighting equipment. Many insurance companies also have inspection services that can be useful for preventing fires in the boiler room.

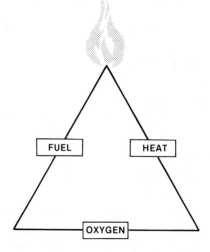

Figure 10-2. A fire requires fuel, heat, and oxygen to burn. The fire will go out when any one is removed.

Classes of Fires

The class of fire is determined by the combustible material burned. The three most common classes of fire are *Class A*, *Class B*, and *Class C*. Class A includes fires that burn wood, paper, textiles, and other ordinary combustible material containing carbon. Class B includes fires that burn oil, gas, grease, paint, or other liquids that convert into gas when heated. Class C include electrical, motor, or transformer fires.

Class D is a rare, specialized class of fires including fires caused by combustible metals such as zirconium, titanium, magnesium, sodium, and potassium. A special powder is applied using a scoop or shovel to put out this class of fire.

The boiler operator must know where every fire extinguisher is located in the boiler room and plant. The boiler operator must also know what type of fire extinguisher is used for each class of fire and how to use different types of fire extinguishers. See Figure 10-3.

Fire extinguishers are not meant to take the place of the local fire department. Fire extinguishers are only meant to put out small fires or help to contain larger ones until additional help arrives.

The number and type of fire extinguishers needed are determined by the authority having jurisdiction and are based on how fast a fire may spread, potential heat intensity, and accessibility to the fire. Additional fire extinguishers must be installed in hazardous areas. The National Fire Protection Association lists these areas as *light hazard* (low), *ordinary hazard* (moderate), and *extra hazard* (high).

Light hazard areas include buildings or rooms that are used as churches, offices, classrooms, and assembly halls. The contents in buildings and rooms of this nature are either noncombustible or not anticipated to be arranged in a manner that would be conducive to the rapid spread of a fire. Class B flammables (for example, fluid for duplicating machines) stored in a light hazard area must be stored in closed containers.

Ordinary hazard areas include shops and related storage facilities, light manufacturing plants, automobile showrooms, and parking garages. In addition, ordinary hazard areas include any location where Class A combustibles and Class B flammables exceed those expected for light hazard areas.

Extra hazard areas are those locations where Class A combustibles and Class B flammables exceed those expected in ordinary hazard areas. Extra hazard areas include woodworking shops, manufacturing plants using painting or dipping, and automotive repair shops.

Type of Agent	Multi-Purpose Dry Chemical (Monoammonium Phosphate)	Regular Dry Chemical (Sodium Bicarbonate)	Halon 1211 (Bromochlorodifluoromethane)	Carbon Dioxide	Water
Each class of fire calls for the right kind of extinguisher. Using the wrong extinguisher is dangerous and may do more harm than good. For your own protection, you should know the classes of fire, the different types of extinguishers, how to use them and why.					
A Fires in ordinary combustible materials—paper, wood, fabrics, rubber and many plastics. Quenching by water or insulating by Multi-Purpose (ABC) dry chemical is effective.	**Yes—excellent** Adheres to burning materials and forms a coating which will smother the fire and minimize reflash.	**No**	**Yes—excellent** Halon 1211 leaves no residue, may not normally affect equipment.	**No**	**Yes** Water saturates material and prevents rekindling.
B Fires in flammable liquids —gasoline, oils, greases tars, paints, lacquers and flammable gases. Multi-Purpose (ABC), Regular Dry Chemical, Halon 1211 and Carbon Dioxide agents smother these fires.	**Yes—excellent** Dry chemical agent smothers fire. Screen of agent shields user from heat.	**Yes—excellent** Dry chemical agent smothers fire. Screen of agent shields user from heat.	**Yes—excellent** Halon 1211 leaves no residue, may not normally affect equipment.	**Yes—excellent** Carbon dioxide leaves no residue, may not normally affect or damage equipment.	**No** Water will spread flammable liquids and not put it out.
C Fires in live electrical equipment—motors, generators, switches and appliances—where a non-conducting extinguishing agent Multi-Purpose (ABC), Regular Dry Chemical, Halon 1211 or Carbon Dioxide is required.	**Yes—excellent** Dry chemical agent is non-conductive. Screen of agent shields user from heat.	**Yes—excellent** Dry chemical agent is non-conductive. Screen of agent shields users from heat.	**Yes—excellent** Halon 1211 is a non-conductor, leaves no residue, may not normally affect or damage electrical equipment.	**Yes—excellent** Carbon dioxide is non-conductive, leaves no residue, may not affect or damage electrical equipment.	**No** Water, a conductor, should never be used on live electrical fires.

Figure 10-3.   Fire extinguishers are classified by the combustible material burned. (*Walter Kidde*)

SAFETY RULES

The boiler operator is the person responsible for the safe and efficient operation of the boiler. The boiler operator must develop safety habits to prevent personal injury, injury to others, and damage to equipment. Safety rules vary depending on the type and size of the plant. However, the basic safety rules listed are common to all boiler rooms.

1. Wear approved clothing and shoes in the plant at all times.
2. Wear gloves when handling hot lines or cleaning fuel oil burner tips.
3. Wear appropriate eye protection in all designated areas. Use hand shields when visually inspecting the furnace fire.
4. Wear goggles and respirators when cleaning the fire side of the boiler, breeching, or chimney.
5. Wear a hard hat when working where there is a possibility of head injury.
6. Do not use hands to stop moving equipment.
7. Store all oily rags or waste in approved containers to prevent fires caused by spontaneous combustion.
8. Only use approved safety cans to store combustible liquids.
9. Check all fire safety equipment on a regular basis to be sure it is in proper working condition.
10. Check fire extinguishers periodically for proper charge and correct location.
11. Do not use unsafe ladders or substitutes for ladders.
12. Ladders should never be used as bridges.
13. Do not leave loose tools on ladders, catwalks, tops of boilers, or scaffolds.
14. Do not carry tools in back pockets.
15. Do not throw a tool to anyone at any time.
16. Use the proper tool for the job.

17. Do not use defective tools.

18. Always secure and tag steam stop valves, bottom blowdown valves, and feedwater valves when a boiler in battery is removed from service for cleaning and inspection.

19. Never start any equipment that has been tagged out for safety reasons.

20. Make sure the equipment has been secured and tagged out before attempting to clean or repair.

21. Always use low voltage droplights when working in boiler steam and water drums.

22. Precheck all equipment for starting hazards.

23. Clean up liquid spills at once.

24. Move quickly and with purpose in emergencies but do not run.

25. Personally double check the plant and equipment before starting up or making repairs. Do not take any boiler function for granted.

26. Make repairs on live equipment only in extreme emergencies.

27. Always report any unsafe condition in the plant to the immediate superior.

28. Only authorized personnel should repair electrical equipment.

29. Never remove high voltage fuses without proper tools.

30. Only use approved solvents for cleaning.

31. Always stand clear when hand testing a safety valve.

32. Boiler valves should be opened and closed slowly to prevent water hammer and damage to pipes.

33. Open the boiler vent to prevent a vacuum from forming before opening manhole and handhole covers.

34. Always check for proper venting when draining a boiler.

35. A pressure vessel that contained a toxic agent

should not be entered without proper precautions.

36. Never enter an enclosed area that has toxic or flammable gas present without adequate ventilation.

37. Always determine the water level in the boiler before removing a handhole or manhole cover.

38. When blowing down, the quick-opening valve should be opened first and closed last.

39. Automatic combustion controls should be checked routinely for proper operation.

40. Never operate any boiler equipment above its rated capacity.

41. Never engage in horseplay.

Points to Remember

1. Report all accidents and have all injuries treated no matter how small.
2. Know the location and types of fire extinguishers required for different types of fires.
3. A clean boiler room will have fewer accidents because many causes will be removed.
4. Learn the procedure in a specific plant for calling the fire department or sounding a fire alarm.
5. Practice good safety habits in all boiler operations. Study safety rules and learn how to put them into practice.

Key Words

classes of fires
combustible material
extra hazard
fire extinguisher
fire prevention

flash fire
light hazard
ordinary hazard
spontaneous combustion

Glossary

A

air atomizing burner. Type of fuel oil burner in which pressurized air atomizes fuel oil in the burner nozzle.

air control. Boiler fitting that removes air from the forced circulation hot water system.

air to fuel ratio. Amount of air and fuel supplied to the burner.

air vent. Allows the release of trapped air in heating units of a hot water heating system.

anthracite coal. Hard coal that has a high fixed carbon content.

aquastat (limit control). Boiler fitting that controls the starting and stopping of the burner by sensing the temperature of the water in a hot water boiler.

ASME code. Code written by the American Society of Mechanical Engineers that controls the construction, repairs, and operation of steam boilers and their related equipment.

atmospheric pressure. Pressure at sea level (14.7 psi).

atomize. To break up liquid into a fine mist.

aquastat

automatic city water makeup feeder. Used to add city water directly to the boiler in the event of failure of condensate returned to the boiler.

auxiliaries. Equipment necessary for the operation of a boiler.

B

baffles. Direct the path of the gases of combustion so the maximum amount of heat is absorbed by the water before the gases of combustion enter the breeching and chimney.

bituminous coal. Soft coal that has a high volatile content.

blowdown valve. Valve located on the lowest part of the boiler opened during blowdown.

boiler explosion. Caused by a sudden drop in pressure (failure on the steam side) without a corresponding drop in temperature.

boiler lay-up. Removing a boiler from service for an extended period of time. A boiler can be laid up wet or dry.

boiler room log. Data sheet used to record pressure, temperatures and all other operating conditions of a boiler on a continuous basis.

boiler shutdown. Sequence of operations completed when taking a boiler off-line.

boiler start-up. Sequence of operations completed when preparing a boiler for service.

boiler tubes. Straight or bent tubes used to carry water or heat and gases of combustion.

boiler vent (air cock). Line coming off the highest part of the steam side of the boiler that is used to vent air from the boiler when filling with water and when warming the boiler. Also used to prevent a vacuum from forming when taking the boiler off-line.

bottom blowdown. Process of discharging water from the boiler to control high water, remove sludge and sediment, and regulate chemical concentrations.

bottom blowdown valves. Valves opened during a bottom blowdown.

Bourdon tube. Connected by linkage to a pointer that registers pressure inside a pressure gauge.

breeching. Duct connecting boiler to chimney.

British thermal unit (Btu). Measurement of the quantity of heat necessary to raise the temperature of one pound of water 1 °F.

bypass feeder. Introduces chemicals into feedwater to prevent scale and reduce oxygen in the boiler water.

Bourdon tube

C

calibrate. To compare and adjust a pressure gauge to conform to a test gauge.

carryover. Water carried over into steam lines that leads to water hammer and possible pipe rupture.

cast iron sectional boiler. Type of boiler that has the heat and gases of combustion flowing around large cast iron sections which contain water.

chattering. Rapid opening and closing of a safety valve caused by improperly set blowdown.

check valve. Automatic valve that controls the flow of a liquid in one direction.

chimney. Outlet to the atmosphere for the gases of combustion. Can also be used to create draft.

circulating pump. Centrifugal pump used in the hot water circulation system to circulate hot water.

classes of fires. Class of fire is determined by the combustible material burned. The three classes of fires are: Class A, started from wood, paper, or rags; Class B, started from oil, grease, or flammable liquids; and Class C, which is an electrical fire.

coal fuel system. System in which coal is burned in the furnace to sustain combustion and create heat.

coal hopper. Where coal is stored before it enters the boiler furnace.

coaling over. Spreading coal over the entire furnace surface.

combination forced and induced draft. System in which both forced and induced draft fans are used to create draft.

combination gas/fuel oil burner. Equipped to burn either gas or fuel oil, which permits the operator to switch from one fuel to another.

combustible material. Any material that burns when exposed to oxygen and heat.

combustion. The rapid union of oxygen with an element or compound that results in the release of heat.

combustion chamber. Location in a boiler where fuel and air mix, causing combustion to occur.

combustion control. Automatic device that controls the efficient combustion of fuel by regulating the starting and stopping of of the burner, the high and low fire of the burner, the purging of the furnace, and the supply of primary and secondary air.

complete combustion. Burning of all fuel supplied using the minimum amount of excess air.

compound pressure gauge. Pressure gauge that indicates pressure on one side and vacuum on the other.

compression tank. Boiler fitting that acts as a relief device in the hot water heating system by absorbing the variations in water pressure caused by water temperature changes.

condensate. Steam that has lost its heat and has returned to water.

condensate return tank. Where condensed steam (water) is stored before it is delivered back to the boiler by the feedwater pump.

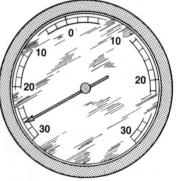

compound pressure gauge

condense. When steam turns back to water after heat is removed.

cross tee. Connection on a water column used for inspection of steam and water lines for scale buildup.

cut-in pressure. Pressure at which the pressure control causes the burner to turn on.

cut-out pressure. Pressure at which the pressure control causes the burner to shut off.

D

damper. Used to control the flow of air or gases through the boiler.

data plate. Plate that must be attached to a safety valve containing data required by the ASME code.

diverter fittings. Used to provide water to a heating unit while maintaining the required flow of hot water to the next heating unit in a hot water heating system.

draft. Difference in pressure between two points that causes air or gases to flow. Draft may be natural or mechanical (forced, induced, combination forced and induced).

draft gauge. Device used to measure draft (the difference in pressure) between the atmosphere and the point of measurement.

draft system. Provides the necessary air for combustion to the boiler.

duplex strainer. Removes solid particles from the fuel oil in a fuel oil system. Allows cleaning of one strainer while the other strainer is in operation.

E

electric fuel oil heater. Used to raise the fuel oil to the proper burning temperature.

erosion. Wearing away of metal surfaces in a boiler.

evaporation test. Test that allows the water level to drop in the boiler to check the operation of the low water fuel cutoff.

excess air. Air above the theoretical amount of air needed for combustion.

expansion tank. Functions as a relief device in the natural circulation hot water system by collecting excess water as the water is heated and its volume increases.

F

fast gauge. Pressure gauge that records a higher pressure than is actually present.

feedwater. Water that is supplied to the steam boiler at the proper temperature and pressure.

feedwater check valve. Automatic valve that prevents water from backing out of the boiler into the feedwater lines.

feedwater heater. Used to heat feedwater and vent air and other non-

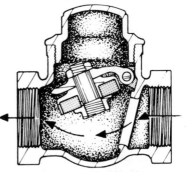

feedwater check valve

condensable gases to the atmosphere.

feedwater lines. Lines going from the feedwater pump to the boiler.

feedwater pump. Takes water from the feedwater heater or condensate return tank and delivers it to the boiler at the proper pressure.

feedwater regulator. Control used to maintain a constant water level in the boiler, which cuts down the danger of high or low water.

feedwater stop valve. Valve that allows the operator to isolate the the boiler from the feedwater line.

feedwater system. Feeds water to the boiler at the proper temperature and pressure.

feedwater treatment. Addition of chemicals to feedwater to protect the boiler from scale and corrosion.

firebox boiler. Type of fire tube boiler that has two different tube lengths and more brickwork than other fire tube boilers.

fire extinguisher. Portable unit used to put out small fires or contain larger fires until the company fire brigade or the fire department arrives.

fire point. The temperature at which fuel oil burns continuously when exposed to an open flame.

fire tube boiler. Boiler which has heat and gases of combustion passing through tubes surrounded by water.

firing rate. Amount of fuel the burner is capable of burning in a given unit of time.

fitting. Trim found on the boiler that is used for safety and/or efficiency.

flame failure. When the flame in a furnace goes out.

flame scanner. Device found on a boiler that proves pilot and main flame.

flash point. Temperature at which fuel oil, when heated, produces a vapor that flashes when exposed to an open flame.

float thermostatic steam trap. Type of nonreturn steam trap in which a float opens and closes the discharge valve according to the amount of condensate in the bowl of the trap.

flow control valve. Prevents natural circulation of water when hot water is not being pumped through the forced circulation hot water system.

foaming. Rapid fluctuation of water level caused by impurities on the surface of the boiler water that can lead to priming or carryover.

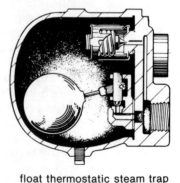

float thermostatic steam trap

forced circulation hot water system. System that uses circulating pumps to circulate hot water.

forced draft. Mechanical draft produced by power-driven fans supplying air to the furnace.

fuel oil burner. Used to atomize fuel oil for efficient combustion.

fuel oil grade. Standard for classifying and numbering fuel oil according to its Btu content which are established by the American Society for Testing Materials.

fuel oil heater. Used to heat fuel oil so it can be pumped and is at the correct temperature for burning.

fuel oil pump. Takes the fuel oil from the fuel oil tank and delivers it to the burner under pressure.

fuel oil system. System in which fuel oil is burned in the furnace to sustain combustion and create heat.

fuel oil thermometer. Found on various parts of the fuel oil system to indicate the temperature of the fuel oil.

fuel system. Supplies fuel in the proper amount, which is burned in order to turn water to steam in a boiler.

furnace explosion. Occurs when fuel oil vapor or combustible gas ignites in the fire side of the boiler.

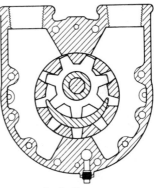

fuel oil pump

fusible plug. Plug that contains a tin center which melts at approximately 450 °F. Used as a last warning device of a low water condition.

G

gas burner. Supplies the proper mixture of air and gas to the furnace so complete combustion can be achieved.

gas cock. Manual quick-closing shutoff valve.

gases of combustion. Gases produced by the combustion process in a boiler.

gas fuel system. System in which gas is burned in the furnace to sustain combustion and create heat.

gas pressure regulator. Used to supply gas to the burner at the required pressure needed for combustion of the gas.

gate valve. Used wherever a direct flow through a valve with no drop in pressure is required. Must be wide open or fully closed.

gauge glass. Located on the water column and used to indicate how much water is in the boiler.

gauge glass blowdown valve. Used to check the water level in the boiler and used to remove any sludge and sediment from gauge glass lines.

globe valve. Found on lines in the boiler room where a throttling action is required. Used to take a piece of equipment out of service for maintenance.

grates. Where the combustion process starts in a coal-fired furnace.

H

hand firing. Manually shoveling coal into the furnace.

hard water. Water that contains large quantities of minerals and scale-forming salts.

heating surface. Part of the boiler where heat transfer occurs. Heat and gases of combustion are on one side and water is on the other.

heating unit. Heat exchanger such as a radiator that extracts heat from hot water or steam and transfers the heat to the air.

heating value. Expressed in Btu's per gallon or per pound. Varies with the type of fuel used.

heat transfer. Movement of heat from one substance to another that can be accomplished by radiant conduction or convection.

high fire. Point of firing cycle when the burner is burning the maximum amount of fuel per unit of time.

high water. When water starts to exceed the NOWL. This condition is dangerous because it could lead to carryover.

hot water boiler. High or low pressure boiler similar in design and construction to steam boilers that uses water instead of steam to transport heat energy.

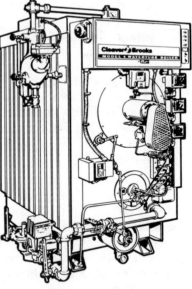

hot water boiler

hot water heating system. Provides heat to designated areas by using water to transport heat energy generated in the boiler.

huddling chamber. Part on a safety valve that increases the area of the safety valve disc, thus increasing the total upward force, causing the valve to pop open.

hydrostatic pressure. Water pressure per vertical foot (.433) exerted at the base of a column of water.

hydrostatic test. Water pressure test on a boiler to check for leaks after repair work.

I

ignition transformer. Used to supply a spark that ignites the gas pilot, which in turn ignites the fuel oil.

incomplete combustion. Occurs when all fuel is not burned, resulting in soot and smoke forming.

induced draft. Mechanical draft produced by a power-driven fan located between the boiler and the chimney.

inverted bucket steam trap. Type of nonreturn steam trap in which steam enters from the bottom and flows into an inverted bucket.

L

lighting off. Initial ignition of the fuel.

live steam. Steam that leaves the boiler directly without having its pressure reduced during process operations.

low fire. Point of the firing cycle where the burner is burning the minimum amount of fuel per unit of time.

low pressure steam boiler. Boilers that operate at a steam pressure not exceeding 15 psi.

low water. Whenever the water level in the boiler is below the NOWL as indicated by the water level in the gauge glass.

low water fuel cutoff. Device located slightly below the NOWL

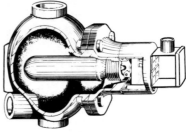

low water fuel cutoff

that shuts off the boiler burner in the event of low water, preventing tube burnout and possible boiler explosion.

M

main steam stop valve. Usually an os&y gate valve used to cut the boiler in on the line or off-line. Located on the main steam line leaving the boiler.

makeup water. Water that must be added to the boiler to make up for leaks in the system, water that is lost through boiler blow-downs, or condensate that is dumped because of contamination.

manometer. Type of draft gauge used to measure boiler draft.

manual city water makeup system. System where the operator controls the water level in the boiler by opening the manual city water makeup valve to allow city water to flow directly into the boiler.

master controller. Receives and processes signals indicating the temperature from a sensing device outside of the building, controlling the burner in order to raise or lower the boiler water temperature.

MATT. Stands for proper mixture, atomization, temperature, and time to complete combustion, which are all necessary for efficient combustion.

MAWP (maximum allowable working pressure). Determined by the design and construction of the boiler to conform with the ASME code.

mechanical draft. Draft produced by power-driven fans and controlled by the speed of these fans and by the use of inlet and outlet dampers. May be forced or induced draft.

mercury tube. Tube in the pressure control that acts as an electrical switch. Flow of electricity in the tube is controlled by the movement of mercury.

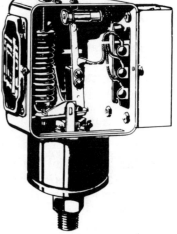

modulating pressure control

modulating pressure control. Sends a signal to the modulating motor that controls the firing rate of the burner. Located at the highest part of the steam side of the boiler.

N

natural circulation hot water system. System that circulates hot water by using the difference in water density of hot water flowing to heating units and cool water flowing to the boiler.

natural draft. Caused by the difference in weight between a column of hot gases of combustion inside the chimney and a column of cold air of the same height outside the chimney.

natural gas. Combustible gas found in pockets trapped underground.

nonadhering sludge. Residue formed in a boiler when feedwater chemicals are added to water that contains scale-forming salts.

nonreturn steam trap. Removes condensate from steam lines and heat exchangers and delivers it to the vacuum tank.

NOWL (normal operating water level). Water level carried in the gauge glass during normal operation (approximately one-third to one-half glass).

O

ON/OFF combustion control. Automatic combustion control that starts and stops the burner on boiler steam pressure demand.

operating range. Range that must be set when using an ON/OFF combustion control which starts and stops the burner on boiler steam pressure demand.

outside stem and yoke (os&y) valve. Shows by the position of the stem whether it is open or closed. Sometimes referred to as a rising stem valve.

oxygen scavenger. Chemical added to boiler water that combines with oxygen to change it into a compound harmless to boiler metal.

P

perfect combustion. Burning of all the fuel with exactly the correct amount of air as accomplished in a laboratory.

pilot. Used to ignite fuel at the proper time in the firing cycle.

pitting. Corrosion of boiler metal caused by high concentrations of oxygen in boiler water.

pour point. Lowest temperature at which fuel oil flows as a liquid.

power-driven fan. Used to create draft mechanically in a boiler.

pressure control. Automatic device attached to the highest part of the steam side of a boiler that controls the operating range of the burner, starting and stopping the burner on steam pressure demand.

pressure gauge. Used to indicate various pressures in the system. Calibrated in psi.

pressure-reducing valve. Reduces incoming city water to approximately 12 to 18 psi for use in a hot water heating system.

primary air. Air supplied to the burner that regulates the rate of combustion.

priming. When small particles of water are being carried over with steam.

programmer. Controls the firing cycle of a burner.

proving pilot. Sighting the pilot through the flame scanner to verify that the pilot is lit.

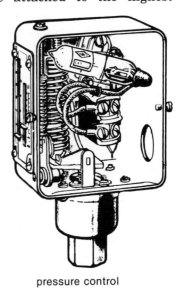

pressure control

psi (pounds per square inch). Unit of measurement used to express the amount of pressure present in a given structure or system.

pulverized coal. Coal that has been pulverized to the consistency of talcum powder and which is highly explosive.

purge. To remove explosive combustibles from the furnace that could lead to a furnace explosion.

R

ram-feed stoker. Uses a feeder block ram to feed coal into the retort of the furnace.

remote temperature-monitoring system. System in which the temperature of the water in the boiler is raised or lowered by controlling the burner, based on the temperature at a remote location.

retort. Space below the grates of an underfeed stoker.

return steam trap. Discharges condensate directly into the boiler when pressure in the trap is equal to or slightly higher than boiler pressure. No longer used on new boilers, but is found on older systems.

rotary cup burner. Type of fuel oil burner that mixes air and fuel by using a fan and rotating cup.

S

safety relief valve. Prevents the hot water boiler from exceeding the MAWP.

safety valve. Prevents the boiler from exceeding its MAWP by relieving excess steam pressure. Found on the highest part of the steam side of the boiler.

safety valve capacity. Amount of pressure a safety valve can discharge per hour.

scale. Deposits caused by minerals in the boiler water.

scale-forming salts. Salts such as calcium carbonate and magnesium carbonate that when in solution tend to form a hard, brittle scale on hot surfaces.

scotch marine boiler. A fire tube boiler with an internal furnace.

screw-feed stoker. Uses a screw to transport coal from the hopper to the retort in the furnace.

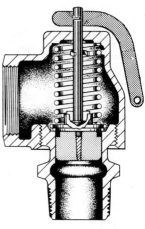

safety valve

screw valve. Valve used as the blowdown valve, taking all the wear and abuse during blowdown.

secondary air. Air supplied to the burner that controls the efficiency of combustion.

siphon. Installed between the boiler and the pressure gauge. Prevents steam from entering the pressure control or the pressure gauge.

slow gauge. Pressure gauge that records a lower pressure than is actually present in the boiler.

sludge. Accumulated residue produced from impurities in boiler water.

sodium sulfite. Type of oxygen scavenger used in boiler water treatment.

soft water. Water that contains none or only small traces of minerals and scale-forming salts.

solenoid valve. Valve that is activated electrically.

soot. Carbon deposits resulting from incomplete combustion that insulate boiler tubes, resulting in high chimney temperatures.

spalling. Hairline cracks in boiler brickwork due to changes in furnace temperature.

spontaneous combustion. Occurs when combustible materials self-ignite.

steam. Vapor that forms when water is heated to its boiling point.

steam boiler. Closed pressure vessel in which water is converted to steam by the application of heat.

steambound. Condition that occurs when the temperature in the vacuum tank or open feedwater heater gets too high and the vacuum pump or the feedwater pump cannot deliver water to the boiler.

steam fuel oil heater. Used to preheat fuel oil before it is sent to the electric fuel oil heater.

steam heating system. Provides heat to designated areas by using steam to transport heat energy generated in the boiler.

steam lines. Used to direct live steam to designated areas.

steam space. Space above the water line in the steam and water drum.

steam strainer. Used to trap scale and impurities that could clog up the steam trap.

steam system. Collects and controls the steam that is created in the boiler.

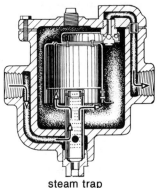

steam trap

steam trap. Automatic device that removes air and condensate from steam lines and heat exchangers.

stoker. Mechanical method of feeding coal into the boiler furnace at a more consistent rate than hand firing.

strap-on thermometer. Used to test a steam trap by indicating the temperature of the steam line before and after the trap.

submaster controller. Receives and processes temperature readings from the master controller and compares them with the temperature of the water circulating in the heating system in a remote temperature-monitoring system.

suction pressure. Pressure on the liquid at the suction side of a pump.

sulfur. Combustible element found in coal and fuel oil.

supply line. Supplies hot water to the heating units in a hot water heating system.

surface blowdown. Removes impurities from the surface of the water that could cause surface tension.

surface blowdown line. Located at the NOWL and used in surface blowdowns.

surface tension. Caused by impurities on top of the water in the steam and water drum of a boiler.

swing check valve. Swings open and closed according to the pressure acting on the valve disc to allow liquid to flow in only one direction.

T

tagged out. Equipment that is being repaired which is marked so it is not operated by mistake.

temperature-indicating crayon (temperature stick). Crayon used to identify temperature of an object. Commonly used to test a steam trap by placing a crayon mark on the discharge side of the steam trap. If the trap malfunctions and allows steam to to pass through, the crayon mark will melt.

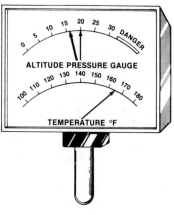

temperature-pressure gauge

temperature-pressure gauge. Indicates the temperature and pressure of the water leaving a hot water boiler.

temperature regulator. Automatic device used to regulate steam flow to various heat exchangers.

thermostatic steam trap. Most common type of steam trap. Flexible bellows, which expand and contract according to whether they are surrounded by steam or condensate, open and close the discharge valve.

three-way mixing valve. Valve in the hot water heating system that automatically blends water returning from the heating units with supply water from the boiler.

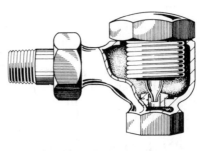

thermostatic steam trap

total force. Total amount of steam pressure acting against the safety valve disc. Total force = area of steam valve disc × steam pressure.

try cock. Usually located on the water column and used as a secondary means of determining the water level in a boiler. With a NOWL, steam should come out of the top try cock, steam and water out of the middle try cock, and water out of the bottom try cock.

U

underfeed stoker. Coal-feeding system that introduces the coal from the hopper into a retort chamber under the fire.

V

vacuum. Pressure below atmospheric pressure measured in inches of mercury (Hg).

vacuum gauge. Pressure gauge used to measure pressure below the atmospheric pressure that is calibrated in inches of mercury.

vacuum pump. Designed to pump both air and water and causes a positive return of condensate to the vacuum tank.

vacuum tank. Collects condensate from heating equipment to be drawn by the vacuum pump to feedwater lines for use in the boiler.

vertical fire tube boiler. One-pass boiler that has fire tubes in a vertical position. Classified as wet-top or dry-top.

viscosity. Measurement of a fluid's internal resistance to flow.

volatile content. Gas content of soft coal.

W

water column. Reduces fluctuation of boiler water to obtain a better reading of the water level in the gauge glass. Located at the NOWL.

water column blowdown valve. Valve on the bottom of the water column used to remove sludge and sediment that collect at the bottom of the water column.

water hammer. Banging caused by steam and water mixing in a steam line.

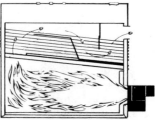

water tube boiler

water tube boiler. Boiler that has water in the tubes with heat and gases of combustion passing around the tubes.

Index